The Harrowsmith Country Life

GUIDE TO WOOD HEAT

The Harrowsmith Country Life
GUIDE TO WOOD HEAT
by Dirk Thomas

Camden House Publishing, Inc.
A division of Telemedia Communications (USA) Inc.

Camden House Publishing, Inc.
Ferry Road
Charlotte, Vermont 05445

Library of Congress Cataloging-in-Publication Data
Harrowsmith country life guide to wood heat / Dirk Thomas
p. cm.
Includes index
ISBN 0-944475-30-2 (s/c) ISBN 0-944475-34-5 (h/c)
1. Fireplaces. 2. Stoves, Wood. 3. Fuelwood. 4. Heating
I. Harrowsmith country life. II Title. III. Title: Guide to wood heat
TH7424.T46 1992
697'.22—dc20 92-13439
CIP

The sections on tree felling in Chapter 4 are adapted from an article appearing in the October 1977 issue of *Blair & Ketchum's Country Journal.* Reprinted by permission of the publisher.

Edited by Castle Freeman
Cover design and illustrations by Carl F. Kirkpatrick
Inside design by Susan Spinelli

Trade distribution by
Firefly Books Ltd.
250 Sparks Avenue
Willowdale, Ontario
Canada M2H 2S4

Printed and bound in Canada by
D.W. Friesen & Sons
Altona, Manitoba

Cover photograph by Craig Blouin

I dedicate this book to my family and friends, among them: my parents, Bruce and Janet Thomas; my wife, Judith; my daughters, Renée and Amy Leah; my friend, Bill de Camp, for his encouragement through the years.

Special thanks to my technical advisor, Daryle Thomas, of the Hearth and Cricket Shop in East Wallingford, Vermont, and to all of the other kind people who took the time to answer my questions.

~

Contents

Chapter 1
Introduction

Woodburning has been a source of fascination for mankind since prehistory. To consider how it has accompanied us on our journey through time is to conjure up colorful and disparate pictures: heavy-browed hunters holding their hands to a warming fire at the mouth of their cave; tipis (or wagons) circled in the prairie night, darkness held at bay by the glowing fire in the center of the ring; the yule log blazing on the hearth of a medieval castle; the glowing cook stove warming the remote New England farmhouse in the depths of an old-fashioned January. Heat is a necessity—it's also a talisman against the dark, a symbol of home, safety, celebration and self-reliance.

It is true that during a period of recent history, wood heating dwindled in most industrialized nations. You'll find few ads for woodstoves in the popular periodicals of the '50s and '60s. Rural folk still burned wood, of course, but it took the Arab oil boycott of the early '70s to reacquaint many of us with our almost lost heritage. The woodburning revival that followed, fueled as it was by soaring oil prices and apparently dwindling supplies of the petroleum products we had taken for granted, had an air of desperation: It was as though we were collectively faced with the task of reinventing the wheel on the very day that we needed to use the wagon. Mistakes were made: Houses designed for central heating were now using wood-fired space heaters; people whose only knowledge of home heating consisted of turning up the thermostat when the weather got cold were quite suddenly attempting to cut their own wood and install and operate wood heaters; woodstoves were being designed and built by people whose only qualifications were ownership of a welding torch and a grasp of the principle that bigger is better.

The results were predictable: A lot of trees that shouldn't have been were cut and burned; a lot of chimneys puffed a lot of particle-laden smoke and suffered chimney fires; some houses burned and some people got injured or killed in the fires or in woodcutting accidents. But also, millions of North American households freed themselves partially or entirely from dependence upon imported petroleum products for residential heat, and a new industry was born: the loose infrastructure of stove and fireplace dealers, manufacturers and installers; researchers; chimney sweeps; and firewood producers that serves today's woodburner.

As the '70s gave way to the '80s and oil prices lost their shock value, a certain number of people returned to the convenience of oil, gas and electric heat. A number of others, however, having discovered that they liked the savings, the nature of the heat, the independence or all three, kept on burning wood. (Exact figures are impossible to find, but estimates of the number of U.S. households using wood for all or part of their heat range as high as 25 million.)

Thanks to the nearly twenty years of research spurred by the woodburning revival in the '70s, we know a great deal more than we did about wood, stoves, chimneys and wood heat in general. Thanks in part to Environmental Protection Agency regulations, we have fewer stove manufacturers than in the '70s, but the stoves they produce are far cleaner burning and more effective than were the airtight boxes of two decades ago. Nevertheless, the future of wood heat is largely dependent upon two questions that must be conclusively answered, probably before the '90s draw to a close: 1.) Is wood heat economically viable? And, most important, 2.) Is wood heat environmentally defensible?

The final answers—as final as answers to this sort of question can be—will come partly from researchers and manufacturers, but it is, in large part, the individual woodburner who holds the key: Will we be educated and responsible enough to make full use of technological advances in wood heating? Or will we cut the wrong trees and burn their wood improperly in the wrong stoves until laws springing from environmental necessity force us to abandon our woodpiles in favor of the oil tank, the gas pipeline, the electric meter?

Bearing in mind that wood heat and people who burn wood don't

lend themselves very well to generalizations, let us attempt to find preliminary answers to our two questions.

Is Wood Heat Economically Viable?

I'm reminded of a tongue-in-cheek article I saw in the '70s that purported to be an accounting of the "real" expenses involved in cutting your own winter's supply of firewood. Included were a $10,000 truck, chain saws, trips to the hospital emergency room and much lost time at work. Needless to say, the cost per cord was prohibitive. There can be a certain amount of truth to this gloomy view of cutting your own wood; for some people, it's a good way to save money, but it's not for every woodburner, particularly since wood can be purchased cut, split and delivered in most areas at prices that make it a bargain when compared with other fuels.

Simple math shows that, other things being equal, you will spend fewer dollars heating your house with wood than you will with oil, gas, coal or electricity. A cord of good hardwood will produce as much heat as will 200 gallons of heating oil, and the comparison with other fuels is similar. What this means, in dollars and cents, is that if oil costs $1 per gallon, you could pay $200 for a cord of wood—an unheard-of price outside of major cities—and be getting as much heat for your money. So why doesn't everyone use wood? Because, harking back to the apocryphal tale with which we began this discussion, simple math does not address all the factors that the prospective woodburner must consider. Your time, for example. Wood heat is a decidedly active pursuit; even if you buy your wood ready to burn, you'll devote considerable time to piling, carrying, stoking and ash removal. If you cut your own, you'll find yourself warmed more than twice by the felling, bucking, splitting and hauling you'll do, and you'll also find that the time you commit to getting in your wood makes heating your home more of an ongoing hobby than an occasional chore.

Does it make economic sense, then, to heat your home with wood? Yes, if you have more time than money, and yes, if you enjoy the work and ritual unique to this form of heat. Burning wood fits some ways of life in the same way that vegetable gardening and livestock raising do: It saves money, but the savings are almost incidental to the satisfaction it can provide.

Is Wood Heat Environmentally Defensible?

There are two major environmental issues connected to wood heat: air pollution and deforestation.

The air pollution issue is something of a cause celebre in the wood heat industry. Some municipalities now have strictly enforced "no-burn" days and limits on installations of woodburning appliances, and the Environmental Protection Agency has mandated that, as of 1990, no stoves may be manufactured that emit more than 4.1 grams of particulate matter per hour. (To put that figure in perspective, consider that the "black box" stoves typical of the 1970s poured up to 40 grams per hour into the atmosphere.)

Woodsmoke contains carcinogens, so the concern is legitimate. Other environmental alarms about woodsmoke require closer examination. Concerning the "Greenhouse Effect," for example, in which carbon dioxide in woodsmoke degrades the earth's ozone layer leading to global warming, it is worth noting that the carbon dioxide produced by burning wood would have been produced even had the wood not been burned, when the trees that it came from died and decomposed. (This is not true of fossil fuels, however, since they are extracted from below the earth's surface and do not decompose.)

The response of the wood heat industry to environmental concerns is that the new, high-tech stoves, with their drastically reduced emission levels (some models now produce less than 1 gram per hour), will vastly reduce air pollution from woodburning. Also, there is some suspicion in the industry that wood heat is being unfairly singled out as a culprit. A story making the rounds of chimney sweeps recently concerns a community with a no-burn day. The story has it that the inspector whose job it is to drive around and order people to extinguish their fires would have to make a bust every ten or fifteen minutes to prevent more pollution than his car is putting into the atmosphere. This may be true, but it is also true that it is incumbent upon the woodburning public to do its share toward alleviating the potentially catastrophic problems of air pollution and ozone depletion by learning to burn wood properly and by switching to clean-burning stoves whenever possible.

The issue of deforestation does not lend itself to simple answers. While it is true that firewood harvesting can be an excellent tool for encouraging sound forest management, it is equally true that not all firewood is harvested responsibly. It is safe to say, however, that massive deforestation—in tropical and temperate rainforests, for example—is associated with timber production and land clearing, not firewood production. Vermont, as a case in point, is a bastion of woodburning but still grows more wood every year than it cuts.

A last point concerning the ethics of woodburning is that no form of home heating, with the possible exception of solar, is environmentally benign. And the nature of wood as heating fuel makes it less harmful than other fuels in certain ways not immediately related to the burning of the fuel. Consider:

1.) Wood, since it is generally produced locally, requires far less transportation than other heat sources—no oil tankers, pipelines, long-haul trucks, electrical transmission lines, etc.

2.) A spilled load of firewood is an inconvenience, not an ecological disaster.

3.) Responsibly harvested, wood is a renewable energy source.

4.) Wars over firewood have been rare or nonexistent, and there is little reason to believe that this will change. Woodburners are mostly too busy cutting, splitting and piling to do much fighting.

5.) Electricity and natural gas require substantial rights-of-way for transmission. These are, in effect, clearcuts which must be perpetually maintained, often with herbicides.

6.) Neither wood nor its residue—ashes—is a hazardous waste.

I could go on, but I think I've made my point: Every kind of heat, including wood, has an impact on the environment. Is heating with wood ethical? It may not be if you live in an area plagued by air pollution. It probably isn't if you live in an area with little forested land. It isn't if you harvest and burn your wood irresponsibly. If, on the other hand, your circumstances permit and you decide to become a responsible user of the resource, woodburning can be an integral part of a contained and conserving way of life with positive ecological impacts balancing the negative. It is my intention that this book be a useful tool for aiding and encouraging responsible, ethical and enjoyable use of firewood.

Chapter 2

Who Should Burn Wood?

Who should burn wood? This looks like a simple question, but considering the time, money and frustration (not to mention the discomfort) that may attend a wrong answer, it's worth a fairly exhaustive investigation. What we are really asking is not one simple question, but a series of involved ones.

Where do you live? In a valley where atmospheric inversions cause frequent smog? In an area with a relatively warm climate? A place with little nearby woodland? In the introductory chapter we discussed the increasing awareness on the part of the public and certain state and federal regulatory agencies that woodsmoke, for all its evocative and benign associations, is an air pollutant—an important one in some areas. Having already stated that I feel wood heating is getting a somewhat undeserved environmental bum rap born of oversimplification, I must admit that this most ancient and cozy source of heat can have negative impacts.

I remember when it first occurred to me that wood heating was potentially harmful. It was probably 1975 or 1976—the heart of the woodburning revival and the era of the airtight black-box stove—and I was driving into Rutland, Vermont, at dawn. It was a bitterly cold and clear January morning, and as I topped the hill that afforded me my first view of the city, I was shocked to see what appeared to be a deep lake of cloud hovering over the narrow Rutland valley, supported by countless plumes of chimney smoke. Vermont's air then—as it does most of the time now—seemed wonderfully clean and invigorating to me, but this dense cloud was not alien, not a bad-mannered tourist from Newark or Los Angeles; it was our very own, a creation of our innocent desire to warm ourselves with the resources at hand. Oh, I know, car exhaust and oil burner fumes contributed to that cloud, but the distinct and pervasive odor of woodsmoke in Rutland that morning left no doubt about what was to blame.

The smog that I saw that cold morning was partly a result of poorly designed and operated stoves, but mainly a result of the concentration of a large number of those stoves on the floor of a fairly steep-walled valley. It's worth noting that none of the hill towns in the area had smog that morning, even though many of their houses were assuredly heated by poorly designed and operated stoves. The culprit, then, was an atmospheric inversion—cold air at ground level, warm air above, and no mixing or air movement occurring, because cold air doesn't rise. Inversion most often occurs in valleys—particularly steep-walled valleys—because they collect and hold cold air.

The lesson to be taken from this? If you live in a well-populated valley, you'd be well advised to think twice before planning your winter around a woodstove: Besides contributing to a real air pollution problem, you may—now or in the future—find your use of wood heat seriously constrained by local ordinances. At the very least, plan on using a new-generation, low-emission woodstove.

Inversion-prone areas aren't the only places unsuitable for serious wood heating. If you live in a warm climate—winter temperatures seldom going below freezing—you may find that a woodstove causes more problems than it solves. As a chimney sweep in northern New England, I have learned that the dirtiest and therefore most dangerous chimneys belong to people who run their stoves regularly in the spring and autumn. They light a fire to take the chill of a cool rainy day out of the house, but once the stove has accomplished that, it keeps throwing heat, forcing the owner to either shut off its air supply and cool the stove down or go outside and stand in the rain. Woodstoves—both black boxes and low-emission units—don't burn efficiently at very low temperatures; smoldering fires produce large quantities of air pollution and creosote at a cost to the environment and your safety that outweighs any savings in fuel expenses you might realize. If winters where you live aren't cold enough to consistently demand relatively high-temperature burning, you might do better with an alternative heat source.

The local availability of firewood is another factor to consider. Two of the most important tangible advantages that wood has over other heating fuels are that it's relatively cheap and plentiful and—a related factor—it requires relatively low-impact transportation. If you live in an area with few trees—the Great Plains, for example—or in an area whose trees are not suitable for harvest—New York City—wood's advantages as a fuel become purely theoretical. Trucking wood long distances increases the price dramatically and eliminates or decreases one of wood's environmental advantages over, say, oil: low-impact, short-haul transportation. Cutting trees for fuel in areas with little forestland is selfishly short-sighted. This doesn't mean that you'd have to live in the middle of the forest primeval to burn wood with a clear conscience, just that fuel wood should be a by-product of sound local forestry practices. Before you decide to become a serious fuel wood consumer, investigate the availability of wood in your area.

If the topography, climate and sylvan population of your area check out, the next factor to evaluate is your house. Unhappy unions of heaters with homes are commonplace, often troublesome and potentially dangerous. The design and siting of your house may dictate what sort of woodburning appliance—if any—you should use, but before considering this, let's talk a bit about wood heaters and the types of heat that they produce.

The heaters fall into three broad categories: stoves, fireplaces (which aren't normally "serious" heaters) and furnaces. Stoves and fireplaces produce radiant heat, which travels in straight lines and heats solid objects in its path rather than the air. Furnaces, which may be no more than stoves installed in the basement and connected to ductwork or water pipes, as well as some specially designed stoves, produce primarily circulating heat—warming the air or water which transfers its heat as it is circulated. Neither the radiant nor the circulating heater has an intrinsic advantage over the other, but your house may make one preferable.

If your house, for example, is a typical ranch—spread out but not up, with numerous interior walls, the most reasonable way to safely heat it with wood is with a furnace in the basement. On the other hand, if part of your reason for considering wood heat is esthetic, and a stove that you can sit in front of is what you want, then resign yourself to heating part of your home, but not all of it, with wood. There's nothing wrong with this—in fact, many wood heating experts feel that the smaller stoves appropriate for this sort of space heating are easier to operate efficiently than are the appliances large enough to heat a whole house. People do heat entire ranch houses with radiant woodstoves, but this usually means using two or more small stoves or keeping the room where the one large stove is located unbearably hot.

A house designed or renovated for the purpose can be nicely heated with a woodstove. Such a house is probably compact, multistoried to take advantage of rising heat, and with relatively few interior walls to inhibit the effectiveness of the stove's radiant heat. We should note that many such houses are also sited to maximize solar gain. This is a wonderful feature, as I can attest since my house is one of these, but it has an impact on the wood heating operation. To illustrate, I'll use myself as an example. For some years I had a stove with a thermostatically controlled air supply; the stove would automatically close its air inlet when the room got warm. This meant that if I guessed the weather wrong and stoked the stove on the morning of a day that became sunny, I'd come home to find that my heating system had shut itself down to a slow burn and thus had been distilling creosote all day. Even a nonthermostatic stove in a solar house can cause problems if you've stoked it full of wood and the sun comes out. It may not shut itself down, but it will surely provoke you to do so, and that, of course, will cause rapid creosote buildup. Open some windows instead, and memorize those fair weather signs.

Is there such a thing as a house which, because of it design, should not be at least partially heated with wood? Perhaps. I'd propose as a candidate a well-insulated, very tightly constructed, one-storied house with no basement (thus making a wood furnace impractical) and no large rooms. In such a house, it would be nearly impossible to burn a stove hot enough for efficiency without overheating the room where it was located.

In all likelihood, your house is neither designed expressly for wood heat nor completely impractical for it, which means that, perhaps with some renovations, you could use wood to fill part or all of your heating needs. In subsequent chapters, we will discuss how in detail.

The final factor to consider in deciding whether you should burn wood is you. Are you willing to take the time to learn how to do it properly and then follow through? Are you willing to deal with wood, ashes, dust and the inconvenience of a source of heat which, though lovely, is not automatic? Wood may save you money, but it won't save you time, and its pleasures make up for its drawbacks only if you feel those pleasures keenly. Many people, of course, heat with wood out of absolute financial necessity, and a number of them, though by no means all, would drop it like a bad habit if they could afford to. For those of you who have the option, I encourage you to choose wood, but only after carefully weighing all the factors.

Chapter 3
Sources of Firewood

Deciding to heat with wood is one thing, but finding a reliable source of fuel can be quite another. The first question to ask yourself is whether you should buy your wood cut, split and delivered or do some or all of the work yourself.

The primary advantage of buying ready-to-burn wood is convenience: Processing firewood is time-consuming, even for skilled and well-equipped professionals. The work also requires a fair amount of physical strength and stamina, has inherent dangers and is best performed with tools and vehicles that can be expensive and that are not found in the typical homeowner's tool chest or driveway. Another advantage is that it is easier to find firewood dealers in many areas than it is to find available standing timber.

The disadvantage of buying ready-to-burn wood is, of course, expense. The dealer must cope with all of the aforementioned problems inherent in wood harvesting just as surely as you would, and his coping has a price. In my neck of the woods, for example, a cord of unseasoned hardwood, cut, split and delivered, will cost $75-$85. The same cord cut, but not split, will cost $60-$65. That cord uncut and unsplit—delivered to your house in random log-lengths of 8 feet to 18 feet—will probably cost $50. The biggest savings by far come when you do the whole job yourself, starting with standing trees, even if you don't own your own woodlot. The stumpage price (the price the woodcutter pays the landowner) for firewood is usually $5 or $6 per cord. Why the big difference in price for wood in its various forms? Let's look at the firewood business.

A friend of mine who sells some firewood ("but not for a living, fortunately") thinks that somebody ought to do an MBA thesis on firewood prices and the market forces that affect them. He says, and I agree, that in our area (and probably in most well-forested areas) the commodity is seriously underpriced, whether you consider its heat value (BTUs per dollar) or the money that an entrepreneur can make per hour selling it. The price of firewood has, in fact, changed little in the past twenty years, while the price of the tools and fuel needed to harvest it has inflated along with everything else. In 1974 my top-of-the-line chain saw cost $290, gas cost 50 cents per gallon and firewood cost $65-$75 per green cord, cut, split and delivered. The top-of-the-line saw I own today cost $750, gas costs $1.35 per gallon, and firewood costs $75-$85 per cord, cut, split and delivered. Why? And why can you save $70-$80 per cord by performing the whole harvesting operation, but only $20 per cord by splitting it yourself?

The answers are competition and equipment. In forested areas there tend to be a fairly large number of people who sell firewood, whether on a full-time basis or part-time as a by-product of logging, which also tends to be a common activity in such areas. This competition keeps the top price relatively low, particularly because loggers and other cutters who don't depend upon firewood as their main salable commodity will often undersell everyone who does and thus force the price down. Another factor keeping the price low is that if current wood buyers had to pay a per-cord price that realistically reflected wood's value, many would switch to alternative fuels which, though more expensive, require little labor of the consumer.

The equipment factor explains why the savings per cord aren't enormous until you do all the work yourself. Chain saws and splitters don't cost a fortune, but skidders and loader-equipped log trucks do, and that's the equipment it took to make that six-cord pile of log-length firewood appear at the end of your driveway. That's why it cost $50 per cord uncut and unsplit. You still have to perform the most arduous and time-consuming labor to get the wood ready for your stove or fireplace, but you won't need $150,000 worth of machinery to do it.

Perhaps the best way to help you decide which wood procurement method best suits you is to describe four scenarios including comparative cost and time estimates. We'll assume that all four scenarios are at least theoretically available to you. (Clearly, city dwellers would have a hard time arranging for delivery of log truck loads of uncut firewood, and an even harder time convincing their neighbors not to complain about the chain saw noise. Clearly, too, many people, for various reasons, cannot consider cutting standing trees.) The prices for green wood that I use are fairly accurate in the rural Northeast; if they are higher or lower where you live, use the figures that are correct for your area. Also, the times given for the tasks described represent an enormous variable: A logger with a 90cc chain saw bucking large-diameter wood into 4-foot lengths can make a cord faster than you can drink a hot cup of coffee. A weekend cutter with a dull 35cc saw bucking small-diameter wood into 14-inch lengths will take a lot longer—almost forever. Similarly, splitting a cord of straight-grained ash might take less than the hour and a half that I allow; splitting a cord of American elm (especially by hand) will likely take much longer, if you're desperate enough to do it. Keeping these variables in mind, let's survey our four scenarios.

1.) *You buy your wood cut, split and delivered.* The cost will be $75-$85 per cord. The only labor left for you to perform is piling the wood so that it will season properly. This job requires no tools and will probably take an hour or less per cord unless you have to move the wood some distance from where it's delivered to where it is to be piled, in which case you'll need a wheelbarrow and probably a good deal more than an hour, depending upon how far the wood must be moved. At some point during the wheelbarrowing, you will probably figure out a way of having the place where the wood is delivered next year be the place where it is to be piled.

2.) *You buy your wood cut and delivered, but unsplit.* The cost, if you can find somebody who sells it this way, will probably be $60-$65 per cord. In addition to the labor involved in piling, you will need to do the splitting. The tools you will need might be a $25 splitting maul or a $2,000 hydraulic splitter, depending upon your bank balance and your proclivities. The time involved? A common old-timer claim is that a good worker should be able to split a cord an hour with a splitting maul. Well, maybe there aren't any good workers left, and maybe the truth has enjoyed a bit of embellishment over the years. It's also possible that old stoves and furnaces accepted unsplit chunks more gracefully than do modern stoves, so fewer sticks per cord actually got split in the old days. Possibly, too, with fewer foresters around, more straight-grained, forest-grown trees found their way to the woodpile, making for easier splitting. Splitting a cord of wood may well be a tougher job now than it was in our grandparents' time. In any case it's probably realistic to allow 1½-2 hours per cord if you're splitting by hand and 1½ hours per cord if you're using a hydraulic splitter. Your savings compared to buying your wood cut, split and delivered will be $10-$25 per cord, and the additional time required will be 1½-2 hours, so by splitting your own wood you can give yourself employment that pays you anywhere from $5 to $17 per hour, though I wouldn't count on making the $17 very often.

3.) *You buy a log truck load of uncut, unsplit, delivered firewood.* The price will probably be $300-$375 for the load, and the load should contain 6-8 cords. Being conservative, we'll say that your price per cord will be $50. In addition to splitting and piling this wood, you will have to buck it—that is, cut it to length. For this you will need a chain saw or a bow saw. The bucking time I allow assumes that you are using a chain saw. A bow saw will take much longer. Without getting fixated on the variables (muddy logs dull your chain—you spend an hour filing for every cord you cut; your saw is small and slow; your saw is big and fast, etc.), we'll allow 1-2 hours for bucking a cord of average logs into 18-inch stove lengths. You've worked 2½-4 hours more than you would have if you'd bought cut, split and delivered, and you've saved $25-$35 per cord. Not a great hourly wage, but few firewood cutters get rich, and unlike most jobs, this one pays more in direct proportion to your production as it increases with experience.

4.) *You buy stumpage (the right to cut a certain number of standing trees on someone else's property).* This should cost $5-$10 per cord. In addition to bucking, splitting and piling, you will have to fell, limb, skid and truck your wood. The necessary tools vary depending upon where the trees are, but at a minimum would include a saw and a pickup truck, and at the other extreme would include a good portion of the $150,000 worth of equipment to which we referred earlier. Before calculating your potential savings, let me point out the obvious: If you don't already have a truck, it probably won't pay you to buy one just to haul firewood. You'll wear it out and replace it before it amortizes itself. For that matter, you'll wear yourself out before a decent pickup truck amortizes itself. If you'll bear with me through a little oversimplification, I'll show you what I mean: Say that you save $80 per cord by buying stumpage at $5, and you burn 4 cords per year. Without allowing for your time, equipment and increased wear and tear on your truck, you save $320 per year over the cost of wood cut, split and delivered at $85 per cord. In ten years, that's only $3,200 saved, and you'd have to be lucky indeed to find a $3,200 truck that would give you ten years of service. The truck looks like an even worse bargain when you consider that it will only save you $180 per year (for 4 cords) over the cost of having log truck loads delivered to your home.

If you already own a truck, can you save money by doing the whole firewood operation yourself, soup to nuts? Perhaps. As with any task related to wood harvesting, the variables are numerous: How far from home are the trees? Can you drive up to them, or will you have to haul them to the truck somehow? How far? Are you a reasonably skillful woods worker, or a neophyte? And the list could go on. This past summer, I bought firewood stumpage for $6 per cord in the state forest. The woodlot was 15 minutes from my house, and the trees marked for cutting were mostly within 100 feet of truck access, with many closer. My truck holds a little less than half a cord, split and stacked, with room left for tools, and I was able to fill it in a little less than an hour and a half. Add the driving time and the hour per cord to pile it, and you find that I was processing wood from tree to pile at the rate of five hours per cord. Subtract the $6 stumpage fee, and you find that I was making $10-$13 per hour, before figuring in gas, oil and depreciation on the saws and the truck. I would do it again because I enjoy it, I already have the equipment for the job, and the wood does represent a several-hundred-dollar cash outlay that I don't have to make, but it certainly doesn't represent a windfall savings over having the wood delivered in some form.

There is a good deal of firewood available that is half a mile or more from the nearest access. Even if you could get this wood for free, it would never pay you to carry it to your truck (unless someone bought the movie rights from you) or to buy or hire a skidder to do it. My point is that once you get away from having wood delivered to you, all bets are off; sometimes harvesting your own wood will be worthwhile, particularly if you factor in some intangibles such as fresh air, fun and exercise, but often it will be a losing proposition unless you own a woodlot and have a truck, tractor and/or draft animals.

The math and advice I've just presented seem to point to the wisdom of buying wood fully processed, but having done my duty by making that clear, I will now encourage you, if your temperament, circumstances and physical abilities permit, to consider doing some or all of the processing yourself. Intangibles shouldn't be discounted, and a true understanding of the work that went into your winter's fuel might make its warmth that much more comforting. But do keep a healthy dose of clear-eyed evaluation handy, unless you have more time to spend than you can use.

If you resist rising to this bait, read on: In the next section we'll consider how to buy firewood fully processed.

BUYING FIREWOOD

As a chimney sweep, I get to know my customers' chimneys, stoves and fireplaces and, sometimes, their woodpiles as well. One day this past October, while cleaning a regular client's chimney, I noted a small accumulation of dangerous glazed creosote—new to this chimney and very unwelcome, since it is difficult to remove and burns intensely in a chimney fire.

"It must be that wood I bought," said my client when I gave her my report. "The fellow said it was dry, but it sizzles a lot. You look at the pile and see what you think."

I did look, and found wood which, if not cut yesterday, was surely not seasoned; even in the dim light of the woodshed it was easy to see the fresh color of the sapwood.

Later that same day I had just finished cleaning a first-time customer's chimney when his wood was delivered: three supposedly seasoned cords for $375. Before the truck even pulled into the driveway, I could smell the fresh sap. A quick look at the pile confirmed what my nose had already told me: This expensive "seasoned" wood was freshly split and likely freshly cut as well.

Buying firewood is tricky. The business is unregulated; there are no state or federal inspectors, no trustworthy stamps of approval. The quality and even the quantity of the product are difficult to assess. It's easy enough to measure a neatly stacked cord, but can you recognize 128 cubic feet of wood dumped in a heap at the end of the driveway? Can you tell dry wood from green? Both of my aforementioned customers are relatively knowledgeable woodburners, and both, nevertheless, had to make do with green wood after paying for seasoned. Some firewood dealers are highly competent and scrupulously honest, but some are neither, so let the buyer beware! As a consumer you must be well informed or very lucky if you are going to consistently enjoy the warmth and satisfaction that good wood can provide.

You can avoid the uncertainty of the market by cutting your own wood, but this course substitutes its own problems (time, labor, the need for equipment) for the ones it eliminates and is probably a viable alternative only for those who enjoy it. If you don't find yourself in that relatively select group, you're wise to make a smart shopper of yourself.

Is It Seasoned, Dry, Wet or Green?

Make no mistake, it is better to burn dry, seasoned wood than green, wet wood. We've probably all heard people laud the virtues of mixing green and seasoned woods for a lasting fire, and perhaps it worked well in old, nonairtight stoves. But modern airtight stoves, whether high-efficiency catalytic or 20-year-old boxes, need high-temperature fires. Burning green wood in an airtight will almost certainly result in low heat output, heavy glazed creosote accumulation in the venting system and, possibly, dangerous chimney fires. Wood, as it seasons, doesn't lose creosote; it loses water. When green wood is burned, much of the heat it produces is used to evaporate the excess water instead of warming your house and burning off the tars and creosote produced by the fire.

The words "seasoned" and "dry" are generally used synonymously when describing firewood that has been cut, split and piled for one year. Nevertheless, wood that was cut one year ago and split yesterday shouldn't be called seasoned. Neither should some wood that was cut, split and piled six months ago, even though it may be dry, depending upon where it was piled. What I'm getting at here is that you need to question a dealer closely about his seasoned wood before you buy it. When was the wood cut? When was it bucked? When was it split? Was it piled off the ground? Under cover? Off the ground and under cover are not part of the generally accepted definition of seasoned wood, but wood split six months ago and stored that way may well be drier than wood split a year ago and left on the ground exposed to the elements.

What if the dealer lies to you? Then it pays to be able to recognize wood that is seasoned and wood that is not. None of the following indications are foolproof—radial checking, for example, can occur within a few weeks of cutting—but these, coupled with experience and the slightly jaundiced eye that experience often brings, should enable you to get what you're paying for most of the time.

Signs that Wood Is Seasoned

1.) *Weight:* Seasoned wood is much lighter than green wood of the same species.

2.) *Smell:* Green wood often has a pleasant, sappy aroma. Seasoned wood will smell like wood, but not as strongly.

3.) *Loose bark:* As wood dries, the bark adheres less tenaciously. This does not mean that any wood with its bark firmly in place is green, but out of a cord of seasoned wood, you should notice a modest amount of barkless wood and woodless bark.

4.) *Color:* The sapwood visible on the ends and split sides of wood billets fades as wood seasons. Different species have different shades and colors, obviously, but if your new load of wood is bright and fresh in color rather than dull and subdued, you'll want to take a closer look.

5.) *Radial cracking:* As wood dries, it often develops cracks, or checks, that radiate from the heartwood out toward the bark. As mentioned before, this phenomenon may be evident long before the wood is dry, but if it is not evident at all, this is another reason for a closer look.

6.) *Green cambium:* If you're pretty certain that the wood in question is green, peel a little bark back with a sharp knife and check the cambium, which is the very thin layer between the bark and the sapwood. If it is green, so is the wood.

Buy Green

While it certainly is necessary to burn dry wood, it is not necessary to *buy* dry wood if you are organized and solvent enough to buy your wood before you need to burn it. There are three compelling advantages to this strategy: 1.) If you store it properly yourself, you know that your wood is seasoned; 2.) Green wood typically costs anywhere from $15 to $50 less per cord than seasoned; 3). Seasoned wood is not always available in many areas, particularly as winter approaches, so drying your own could save you a miserable heating season.

Quantity

An even more common complaint among firewood buyers than supposedly seasoned wood being green concerns the cord that is not a cord. Wood is usually sold by the cord, which is a stacked pile measuring 8 by 4 by 4 feet and equaling 128 cubic feet, including an acceptable 30 to 40 cubic feet of air. You might, of course, buy your wood by the face cord (8 feet by 4 feet by whatever length you agree upon), the half cord, or in lots of more than a cord, but in any case, the price of wood is mainly determined by volume, and unless you are paying extra to have the dealer stack the wood or he or she brings it stacked on the truck for your inspection (laudable but unusual), you'll either have to accept the dealer's word regarding quantity or learn to recognize a cord (or fractions and multiples thereof) when it's dumped in a heap. A good, though strenuous and imprecise way of doing this is to stack and measure a cord (or whatever quantity you have delivered), throw it into a pile and familiarize yourself with the look of the pile.

Another strategy for estimating firewood quantity involves evaluating the truck in which the wood is delivered. How much will it hold? Some years ago I ordered two cords of sugar maple. The dealer explained that the wood was not cut to length or split—hence the low price—but was cut to the size he could handle: large-diameter sticks might be 18 inches long, and small-diameter sticks might be 8 feet long. He said he could bring a cord at a time.

"What have you got for a truck?" I asked, knowing that a cord of green wood might weigh 3 tons or more. (A cord of dry hardwood might exceed 2 tons.)

"Half-ton Chevy with built-up sides," he answered.

"We'll see," I said to myself. Besides being skeptical that his truck could actually carry the weight of a green cord of sugar maple, I knew that random length and width wood thrown—not stacked—on the truck would contain an extraordinary volume of air, making it unlikely that the advertised two cords would be for real. To make a long story short, the two loads measured 1¼ cords cut, split and piled. Fortunately, the dealer was honest enough to make up the difference.

A truck will hold significantly more if the wood is stacked. For example, a dumptruck I used to use to deliver wood held one cord dumped on with a front end loader, but 1¼ cords stacked. A truck cargo area that measures 128 cubic feet won't contain a cord unless the wood is stacked or is heaped well above the sideboards.

Quality

Getting the amount of wood that you pay for is certainly important, but don't let concern about volume distract your attention from the reason you're buying wood: its heat value, not the air it will displace in your woodshed. Heat value, or quality, is a function of density, not volume. For example, a cord of shagbark hickory takes up the same space as does a cord of aspen ("popple" if you don't live in town). The hickory, however, weighs about twice as much as the aspen and has about twice as many available BTUs. A half cord of hickory, therefore, is worth as much as a full cord of aspen. Before you can decide which dealer is offering the best price, you obviously need to know what kinds of wood he's selling. "All hardwood" is not a good enough answer, and neither is "mixed hardwood." Apple is hardwood, and so is elm, but a cord of apple is the equivalent of 244 gallons of heating oil, while the elm only provides the BTUs of 176 gallons. "Maple" is not a good enough answer, either, at least in areas that have both sugar maple and soft maple (i.e., any variety of maple *other than* sugar maple and black maple), since sugar maple is the equivalent of up to 27 more gallons of oil per cord than are some soft maples.

In fairness to firewood dealers, I should point out that unless you're paying a premium for quality, a fair cord is a mix of the firewood species commonly found in your area. You shouldn't expect every stick to be the top of the line, but neither should every stick be poor to mediocre unless the price is reduced.

Here again, you will do well to become an informed consumer. It's not as easy to identify wood when it's cut and split as it is when it's part of a standing tree, replete with leaves, buds and twigs, but it can be done. Some years ago a firewood customer ordered two cords of ash. I delivered it: two full cords, and every single stick was ash. After he'd had a chance to inspect it, he called.

"I ordered ash. What you brought is sugar maple, isn't it?" Sugar maple would, in fact, have been a better buy, but I reassured him that his wood was ash, and he apparently burned it happily. It seems possible, though, that I could have brought him aspen, which certainly isn't as good a buy as ash. He was properly suspicious, but not properly informed.

There are several ways of judging the heat values of different types of wood, but for most people, the most useful measure is how many gallons of fuel oil or cubic feet of gas a cord of wood will replace for heating purposes. A chart showing this makes it easy to figure out how much money you can save burning different types of wood: Simply figure how much fuel you use in an average winter, plug in current prices for that kind of fuel and for wood, and compare.

For simplicity's sake, I group woods into categories of low, medium and high heat values and give each category's approximate equivalence to fuel oil and natural gas. The chart is adjusted for woodstoves burning at 50-60% efficiency, oil burners burning at 70% efficiency and gas furnaces at 80%. Newer units of all three types may be more efficient, but you should still be able to make a reasonable comparison.

Books, knowledgeable friends and experience will help you master firewood identification. You'll still be fooled occasionally; even experts are. (An acquaintance of mine once successfully passed red elm off as ash to a sawmill.) But if you know your trees, you'll be much better able to get your money's worth when buying firewood.

Heating your house satisfactorily with wood requires more work and knowledge on your part than does any other common type of residential heating. Few are the consumers discerning enough to notice a qualitative difference between one tank of oil or gas and another, but equally few are the consumers who wouldn't notice the difference in performance between good wood and bad. The challenge is to learn enough so that you can identify good wood before you try to burn it and before you pay for it. If you care about these things and enjoy working toward self-reliance, becoming knowledgeable should prove to be a pleasure in and of itself, a pleasure at least matched by the reliable warmth and comfort of your good wood.

STORING AND SEASONING FIREWOOD

Whether you've paid a premium for cut, split and delivered firewood, or you've labored long and hard to cut, split and deliver it to yourself, you still have a critically important task ahead of you: storing your wood. A pile of firewood is full of potential heat which may be realized when you feed it to the flames in your stove, furnace or fireplace. That potential may also go unrealized—at least in part—if you neglect the relatively simple chore of properly piling and protect-

HEAT VALUES OF 24 WOOD SPECIES

Woods Having High Heat Value
(1 cord=200-250 gal. of fuel oil or 250-300 cu. ft. of natural gas)

Hickory	Beech
Oak	Yellow Birch
Ash	Hornbeam
Sugar Maple	Apple

Woods Having Medium Heat Value
(1 cord=150-200 gal. of fuel oil or 200-250 cu. ft. of natural gas)

White Birch	Douglas Fir
Red Maple	Eastern Larch
Big Leaf Maple	Elm

Woods Having Low Heat Value
(1 cord=100-150 gal. of fuel oil or 150-200 cu. ft. of natural gas)

Aspen	Red Alder
White Pine	Redwood
Hemlock—Western and Eastern	Sitka Spruce
Western Red Cedar	Cottonwood
	Lodgepole Pine

ing your fuel. I say simple because you don't have to hide your wood from two of the elements—wind and sun—as they will help it dry. Water is the element to guard against, your woodpile's main enemy. Piling your wood under dripping eaves or in an uncovered heap in the yard is comparable to buying or growing a bunch of beautiful fresh asparagus and then overcooking it: You get no refund for ruining it.

Green wood that is piled for seasoning will dry slower than necessary or not at all (depending upon how much moisture it's exposed to) if it isn't protected. Seasoned wood left unprotected will become, for all intents and purposes, unseasoned as it regains the moisture it lost. Don't believe me? Take a stick of bone-dry firewood and submerge it in a tub of water for a few days. Now try to burn it. For that matter, toss a stick of dry wood into some tall grass and then try to burn it after a good two-day rainstorm: It'll be awhile before it produces as much heat as it does sizzle.

Here are three simple rules for good wood storage:

1.) *Allow air circulation.* You can accomplish this by leaving the sides of the woodpile uncovered and by not stacking your wood so tightly that the pile looks like a well-made jigsaw puzzle. If you have the room, piling your wood in long, narrow tiers—one stick wide, ideally—that run north-south to take full advantage of east and west winds will maximize air circulation. Norm Hudson of the Vermont Department of Forests and Parks says unequivocally that siting your woodpile properly for air circulation is the most important factor in good storage. "Pick the best place to hang your laundry—dry and windy," he says, "and that's where to pile your wood. I like my woodpiles one tier wide, as long as they have to be, and as high as my wife can comfortably reach."

2.) *Protect your wood from rain and snow by covering the top of the pile.* You can do this with a tarp or a woodshed roof, but in either case, it's best if the cover isn't resting on the top of the woodpile: Leaving a few inches of air space there abets circulation. The cover or roof should also extend beyond the wood in all directions to keep at least some wind-driven precipitation off the pile. How far the cover extends depends upon how windy it is where you live and how high you stack your wood; the higher the pile, the farther the cover must extend.

3.) *Pile your wood off the ground.* Water attacks from below as well as from above, so if you don't provide something other than bare ground for your wood to rest on, the bottom layer of your pile will stay wet. This may not seem like a big deal, but if your pile is four feet tall, you might be losing 10% or more of your wood to ground moisture. I build my woodpiles on scrap lumber, but I'd use wooden pallets if I could get enough. They're perfect for the job: They keep the wood well off of the ground and, because they're slatted and hollow, they allow air to circulate from the bottom. It's almost too good to be true.

FIGURES 1 and 2 show different types of supported coverings and woodsheds.

A note about location: On the one hand, you'd like to pile your wood as near to the stove as possible. On the other hand, the termites and carpenter ants would probably like that as well, or better, than you do. Pest control experts I've consulted say that the farther from your house you can pile the wood, the better—as much as 150 feet isn't too far, according to at least one, but it's probably too far to

FIGURE 1. *Temporary supported covering*

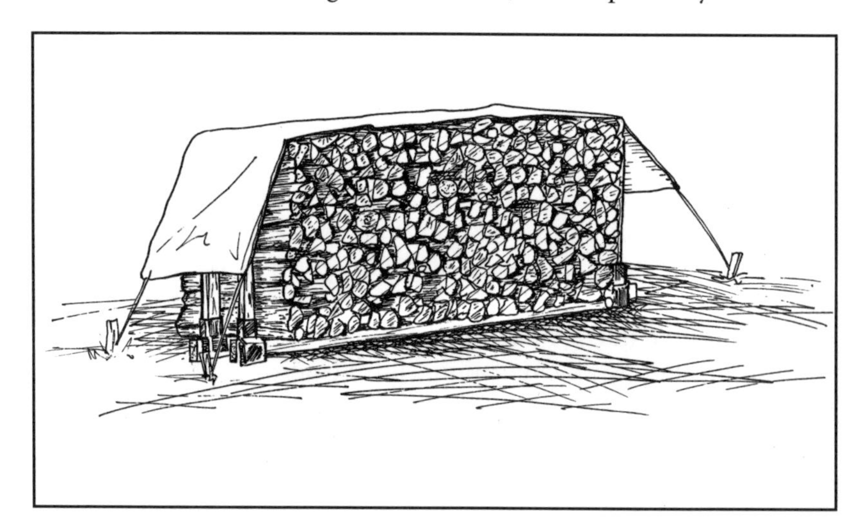

FIGURE 2. *Open woodshed*

carry it. My woodshed is about 50 feet from my house: The shed has carpenter ants, but the house doesn't. I've known people who kept their wood in their basements or in attached woodsheds and had no trouble with insects, and I've known other people who have gotten so tired of hiring exterminators that they've built new woodsheds well removed from their houses. So the location of your woodpile is a compromise between your fear of bugs and your fear of carrying armloads of wood great distances in inclement weather. So much about woodburning seems to be a matter of compromise.

Stacking

Why bother stacking wood? As chores go it's fairly time-consuming, moderately arduous and thoroughly unexciting, and I suppose, with a tip of my hat to the lazy side of the brain, that it wouldn't be necessary if you had unlimited appropriate—covered and off the ground—space in which to toss it. Most people do not have such capacious woodsheds, however, so a good reason for stacking wood is that stacked wood of a certain quantity takes up less space—or, at least, has a smaller footprint—than does the same quantity of wood tossed in a heap. Other reasons are that stacked wood is more receptive to air and sun than is a big, dense pile, and nicely stacked wood looks neat and shipshape, if that matters to you.

How should you stack your wood? Again, this seems simple, but it's surprising how many otherwise capable and intelligent people are unable to stack their wood so that it doesn't keep falling down. Falling down, of course, is the one wrong thing that a stack of wood can do.

Wood stacks tumble for several reasons: 1.) They're not properly supported at the ends; 2.) They're built too high but not straight enough; and 3.) Strong frost heaves topple them. You can support the ends of your wood stacks either by piling against stakes driven into the ground or spiked to the pallets or boards that you're piling on, or by cross-hatching or log-cabin piling the ends of each tier (FIGURE 3).

My woodpiles have a tendency to get crooked and unstable when I pile more than 4 feet high, which I habitually do because of space limitations. This is partly because I'm always in a hurry and, consequently, somewhat inattentive, and partly—I'm trying to let myself off the hook a bit here—because firewood is not uniform, and the higher you pile it, the more you have to compensate for its asymmetry. Nonetheless, I generally get away with building 6- or 7-foot-tall stacks by forcing myself to slow down, check my work, and pile those highly individual sticks of wood so that they complement each other

FIGURE 3.
Cross-hatch the ends of your wood stack for stability.

and produce a stable, symmetrical whole. Although stacking may be unexciting, it does occupy hands and mind well enough to make the time pass quickly.

I've had apparently stable woodpiles tumble in the spring when the frost comes out of the ground. This probably wouldn't happen if I piled on something more substantial than disintegrating scrap wood. Repiling two or three cords of wood that have tumbled together sideways into the tall, wet grass outside the shed is enough of a pain in the neck to make a reasonable person get hold of some pallets.

Seasoning Wood

The time-honored way to season wood is to buck, split and stack it, off the ground and under cover—that phrase again—for one full year. This method will yield firewood with a moisture content of approximately 20%. (Green wood may have a moisture content of 50%.) It's well worth noting that drier is not always better: Wood can be too dry. As we mentioned earlier, wood does not lose any creosote as it dries; it only loses moisture. What would happen if you dried your wood in, say, a greenhouse or even a covered shed for two or three years? Probably the moisture content would dip well below 20%, and the wood would ignite and burn very easily. In an open fireplace or, perhaps, a masonry heater (see Chapter 7), this might be no problem at all. In a woodstove or a furnace, however, this super-dry wood would throw so much heat so quickly that you would shut down the air supply in order to avoid damaging the heating system and cooking yourself. The

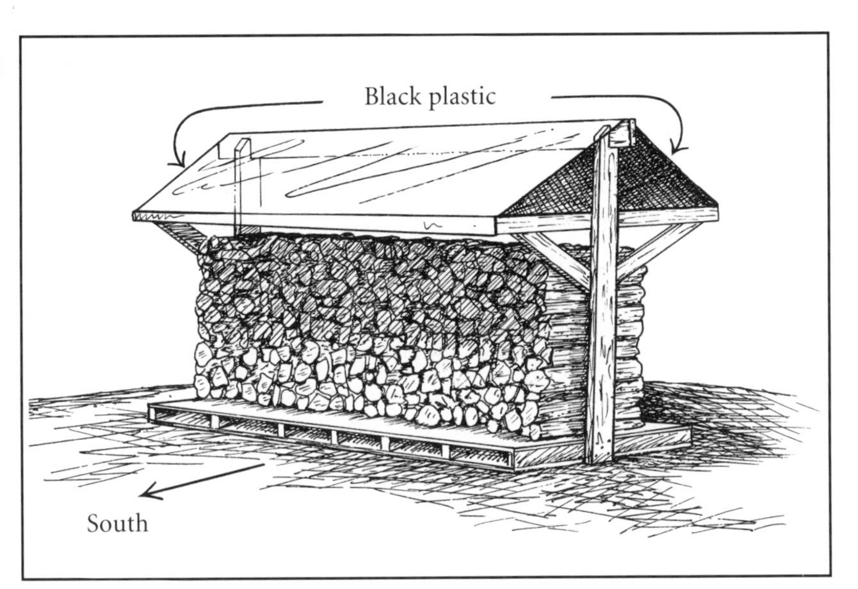

FIGURE 4.
Solar wood drier

product, in the firebox, of high temperatures and little air is smoke—lots of it. Smoke is the medium that carries creosote, and that's just what you'd get. This is why the National Chimney Sweep Guild recommends using wood with a moisture content of 15-20%.

So, if your wood is properly air-dried for one year, you should be all set. But what if you don't have a year? What if you don't get your wood until May or June? You're still all set, provided that you store it properly, because—and here's where the cat jumps out of the bag and mauls a cherished piece of folklore—according to a field study conducted by the U.S. Forest Service, wood properly piled in a dry and windy place will reach a moisture content of 20% in as little as two weeks! If rain falls on it, of course, the moisture content increases. If it's piled in a damp area where little wind reaches it, the wood may never reach 20% moisture content. So we're back to proper storage: If you store your wood properly in the right place, you don't need a year to season it. Of course, there are variables, such as weather, over which you have no control. (What if it's rainy and damp while you're hoping to shortcut the seasoning process?) Plainly, then, planning to get your wood ready a year before you need to use it must still be seen as a good policy, but if conditions are right, you can have burnable wood in far less time.

Other Shortcuts

Over the years people have devised several ways of seasoning wood quickly, at least in theory. In light of the information cited above, it seems clear that you needn't take heroic measures—how much sooner than two weeks hence do you need seasoned wood?—but in case, for some reason, you can't store your wood in a place dry and breezy enough for quick seasoning, here are three popular shortcuts to seasoning:

1.) *The solar wood drier* (FIGURE 4) is a variation on standard proper seasoning. In this case the sun, rather than the wind, is the key element. You stack your wood east-west so that the long side of the pile is facing south, and build a frame over it. The frame should be higher than the stack of wood and extend beyond it. Cover the frame with heavyweight black plastic, securely anchored, and cook the moisture out of your firewood. It's important that the plastic not be touching the wood and that the sides of the stack be open; otherwise the water from the wood will condense on the plastic and drip back onto the wood instead of evaporating or dripping onto the ground. People who have tried the solar wood drier say that they get well-seasoned wood in two or three summer months.

2.) *Fell trees in the spring.* If you make your own firewood from standing trees, you can try cutting them in the spring as the new leaves are popping out and then leaving them, unlimbed and unbucked but severed from the stump, until the new leaves have completely wilted. The theory is that the leaves, unable to draw moisture from the root system, draw it from the trunk of the tree, leaving you with dry wood. Drying wood this way shouldn't take more than a few weeks. The theory is plausible, and I've been told by folks who've tried it that it works. My comments are that it doesn't really save much seasoning time, if any, over properly storing the wood (which you'll need to do, anyway, after you go back and cut it up) and that, while it might work well on some woodlots, on others you'd have an awful mess of tangled trees to sort out when you did go back to buck and limb.

3.) *Kiln-drying firewood,* which is probably not of practical interest for a do-it-yourselfer, is a technique used by a few commercial firewood dealers. This method produces wood with a moisture content of 15-20% in two or three days.

A final note on the subject of seasoned wood: There's no denying that using properly dry wood is important. But, as any knowledgeable chimney sweep will tell you, having a woodburning appliance appropriate for your needs and operating it properly are just as critical, if not more critical, to the safety and success of your heating operation.

Chapter 4

Getting Your Wood In

We're about to begin the section that deals with wood-harvesting techniques, but before we discuss felling, bucking and the rest of the process that changes trees into fuel wood, we need to consider what is truly the most important part of the job: making the best possible use of your woodlot, or woodlot management. This subject is most obviously of interest to those who own forested land and plan to cut wood there, but cutting on someone else's land does not absolve you of the responsibility for doing the job properly: The damage done to woodlands by careless and/or incorrect harvesting practices is long-lasting at best and permanent at worst.

Woodburners, loggers and foresters are fond of saying that wood is a renewable resource and, of course, it can be. But for firewood to be renewable in a practical sense—for a woodlot to provide a sustained yield of fuel while maintaining its vigor—the resource must be managed wisely. Wise management not only provides more wood in the long run than does careless cutting, it also takes into account the other considerable attributes of a healthy forest, such as enhancement of air and water quality, provision of wildlife habitat and various recreational opportunities and, last but not least, the singular beauty of wooded land.

WOODLOT MANAGEMENT

We have, I trust, established the need to manage your woodlot. What remains is to tell you how, and I can't do that directly. The subject is complex: The people who devise forest-management plans professionally—foresters—have college degrees in the field. It is quite possible for an experienced lay person to make reasonable decisions concerning which trees to cut in a woodlot, but fitting such decisions into an overview that includes fauna and other flora, soil conditions and water resources, possible future market considerations and the goals of the owner is likely a job for a forester. It doesn't take magic to plan a woodlot's future, but it does take a lot of knowledge.

The first key to woodlot management, then, is to get the assistance of a forester, and the first place to look is the nearest office of your state or provincial department of forestry; a public forester may be able to provide the management services you need, but if not, he or she will almost certainly be able to guide you to a qualified private forester.

What a Forester Will Do

A forester will look at your woodlot and make a rough inventory of it: Is there any quantity of wood with present or future market value? Are there any extraordinary problems with insects or diseases that need to be addressed? Does the woodlot include any important wildlife habitat or food sources? Are there bodies of water that need protection? Can access be established? What erosion-control measures will be necessary?

At this point, if there are trees appropriate for harvest, the forester will mark them and designate any roads that need to be established. If your woodlot has enough commercially valuable timber to justify a timber sale, the forester will, if you wish—and you should—manage the sale for a small percentage of the profit. Under this arrangement, the forester works out a contract between you and the logger and provides enough supervision to assure that the job is done as specified.

If, as is often the case, previous management (or lack of it) has left your woodlot with no significant quantity of trees with commercial value as timber, a good forester will advise you of your alternatives: Depending upon such factors as the size of the woodlot, its accessibility, and the tree species that currently stock it, you might be able to manage it for production of valuable timber in the future. Or perhaps the best use of the woodlot is as a source of fuel, recreation, wildlife habitat or a combination of some or all of these possibilities. This is where your wants as the owner of the woodland are crucial.

What You Should Do

As capable as a good forester is, he or she can't read your mind, so your first step—the second key to sound woodlot management—is to figure out what you want from your woodlot. Timber? Firewood? Recreation? Do you want it to provide wildlife habitat? Not all goals can be realized in every woodlot. If, for example, you only have an acre of forested land, it's unlikely that it will provide significant amounts of salable timber, but regardless of your land's limitations, it will help a forester work with you if you have a wish list. If you have your heart set on creating a cross-country ski trail and supplying your firewood needs, this will lead to different management decisions than would an ardent desire to establish a deer wintering yard.

Of course, your wishes and the possibilities inherent in your woodlot may be a poor match, so keeping an open mind is a good idea, but having some notion of what you want is a good idea, too; it will provide your forester with a starting point and, unless timber production is your goal, will let him or her know that you want more than the marking of saw logs. It is probably wise to have your goals in mind when interviewing prospective foresters: Just as it is important for you to keep an open mind when considering the management of your woodlot, it is important that the forester who works with you have an open mind. Someone intent upon managing all woodlots for timber production may not give you the guidance that you and your woodlot need.

I alluded earlier to the responsibilities of those who cut wood on other people's land. I believe that these responsibilities extend beyond waterbarring steep roads and picking up empty oil cans. Again, the resource we're using is renewable, but only if care is used in its exploitation. Encouraging the landowner to seek a forester's services before allowing you to cut may pose an inconvenience to you, but—in my opinion—it represents the sort of care we're going to have to learn to take with all our resources unless, of course, we believe that we can get along without them.

CUTTING DOWN TREES

We discussed in Chapter 3 the different ways of acquiring wood to burn. While for many readers, perhaps most, buying fuel wood cut, split and delivered is the most sensible option, there are many people who, because of financial limitations, ready access to standing trees or a strong inclination to work in the woods, will want to give wood harvesting a try. I can certainly sympathize, for even after nearly twenty years as a professional woodcutter in one guise or other, I know of no work (and, indeed, few forms of play) that I enjoy more, particularly when the air is clear and cool and the woods free of blackflies and mosquitoes. Tree cutting is fine exercise for body and mind and can be soothing to the soul, as well, though sometimes retrospectively. By doing it yourself you do save money, and by doing it properly and responsibly, you can be assured that your fuel is a product of sound environmental management. So, with a nod to those who sensibly choose to abstain from this sometimes trying chore, and with a tip of the hat to those who, sensibly or not, want to give it a try, we'll discuss tree harvesting, beginning with felling, that tricky art which, of necessity, precedes limbing, skidding, bucking and splitting.

Rules of Felling

A guide to tree felling is a little like a guide to truck driving: It can help, but by itself it won't provide the knowledge to do the job. Practice—under the eye of an experienced person—is what is necessary.

The first cardinal rule of tree felling is that each tree is different. If you've cut one, you decidedly have not cut them all. A difference in the grain, in the branches, or in the health of the tree can make it fall very differently from another tree. Assume nothing.

Cardinal rule two is that trees can kill. When you're felling a tree, leave yourself at least two clear escape routes; one isn't enough, since the tree might misbehave, making your one path to safety very undesirable.

The value of escape routes was brought home to me once when I was cutting a tall beech on a steep, snowcovered hillside. The tree appeared to lean downhill, and since that was where I wanted to drop it, I didn't bother to check my possible emergency exits. Well, the beech had a rotten spot on one side, and just as I was finishing my backcut, a strong gust of wind caught the top of the tree, twisting it toward the rotten side, causing it to turn on the stump and fall sideways. I knew just before it fell that something had gone wrong, and I started running faster than I would have thought possible. I hadn't gone far when my neglect caught up with me, and I tripped over a branch in the snow. The beech was headed right for me, and I could hear it over my shoulder. Fortunately, it got hung up a couple of feet above me in a maple, leaving me unscathed. Not everyone is so lucky, and you can be certain that I've checked my escape routes ever since and that I don't start running until I know where the tree is falling.

The third cardinal rule of felling is to check the top of the tree for dead wood—widowmakers. Then check the trees it might hit when it falls. When a tree falls, branches from it or from trees it hits can break off and fly in your direction. A friend of mine got his nose broken by a stick that came out of a tree he was felling.

Cardinal rule four is self-explanatory: Be certain that no one is in the way or within reach of the tree you are felling.

Cardinal rule five: Don't stand directly behind or near the tree when it falls, even if it is heading in the right direction. I've seen trees split and jump 10 or 15 feet straight back from their stumps, and trees can move a lot faster than you or I can.

A final cardinal rule is Don't work alone. I know of a logger working alone whose back was broken when a tree came back at him. It was winter, and he died of exposure.

If these major safety points have scared you, I've accomplished my purpose. Now, let's discuss tools and techniques.

Tools

Trees can be felled with axes, hand saws or chain saws, or axes and saws in combination. Axes are fun to use, but the average person will be able to fell trees more accurately and safely with a saw. It's easier to control your cut with a saw, and you can use felling wedges with a saw. If you use a bow saw or a two-man crosscut, you may find that an axe is useful for chopping out the face cut, but otherwise, stick with the saw.

In addition to the saw you will need two or more (depending on the size of the timber that you are taking on) felling wedges, preferably plastic so that you don't have to file your chain every time you nick one. I like the wedges with raised cleats on one side: They eliminate the tendency of smooth plastic wedges to pop out of the backcut in cold weather, usually at the most inopportune moment. I like to have several sizes handy—smaller wedges for smaller trees, larger wedges for larger trees. In a pinch you can always cut a wooden wedge out of a handy branch.

You will also need something with which to drive the wedges, either a small sledgehammer or the blunt side of a single-bit axe. Some purists maintain that using an axe to drive wedges will ruin it, but I have done it myself for some time now with no ill effects to myself or the axe. I prefer the axe because it gives me two tools on one handle—a hammer and a cutting edge, should I get my chain saw bound in a cut. A 3- or 3½-pound head on a fairly short handle seems a good combination, and it is lighter than a sledgehammer, which is another good thing to be said for the axe. If you are still offended by the whole idea, go ahead and use a sledge. But bring the axe along.

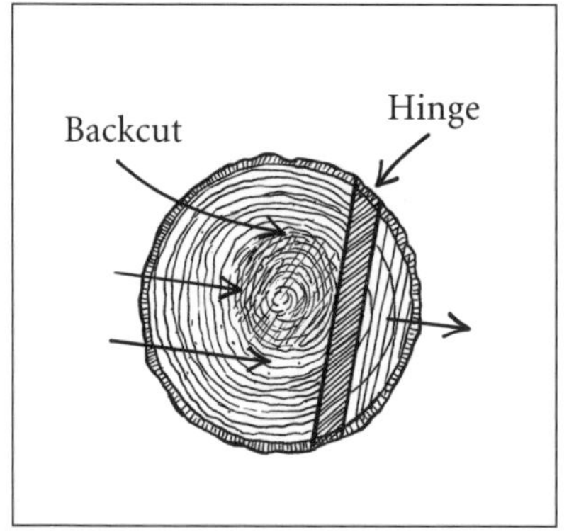

Figure 1. *A hinge as it would look from the top if you had X-ray vision*

Figure 2. *Side view of hinge*

Figure 3. *Front view of undercut*

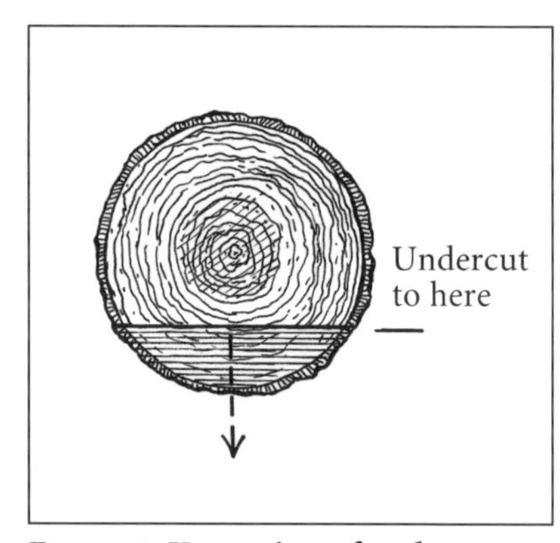

Figure 4. *X-ray view of undercut*

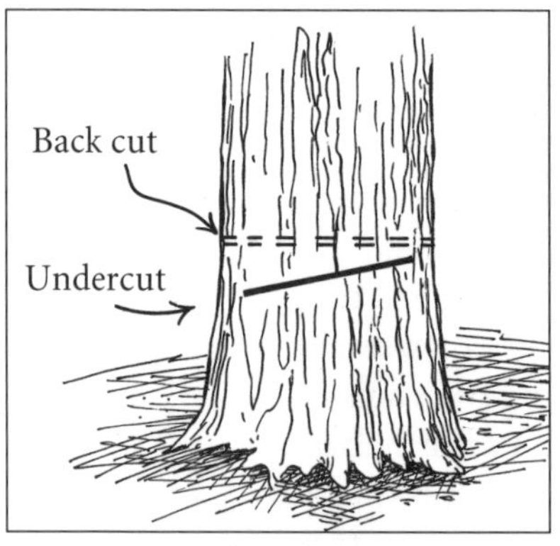

Figure 5. *Avoid angling the under-cut...*

Figure 6. *...since doing so leaves more wood on one side of the hinge than the other.*

FIGURE 7. ***Make the undercut straight into the tree.***

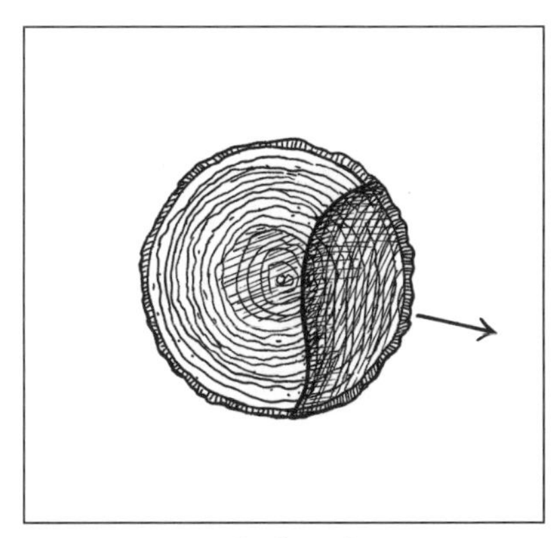

FIGURE 8. ***Crooked undercut***

FIGURE 9. ***Avoid undercutting with the tip of the bar.***

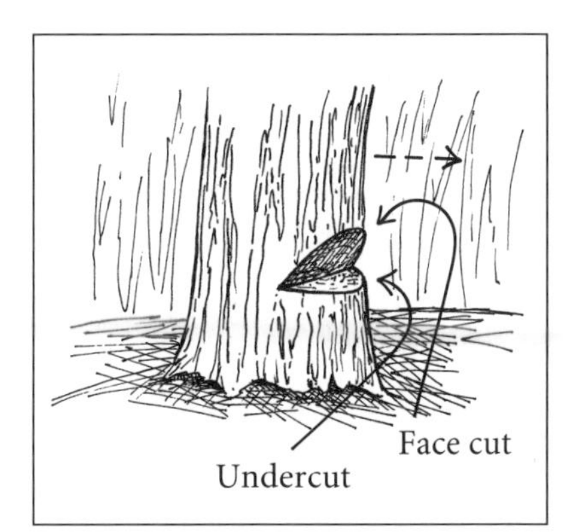

FIGURE 10. ***The face cut***

A saw, axe and wedges should be all you need to fell a tree, unless something goes wrong. But we'll consider that later.

Techniques

A tree is felled by making a series of cuts. One cut would do, except that when you fell a tree with a single cut, you risk splitting the tree and surrendering much of your control over where it falls.

The purpose of the series of cuts is to make a hinge of uncut wood in the tree. When the tree falls, the hinge acts in much the same way as a door hinge and becomes, in fact, your primary tool for aiming the tree. Roughly speaking, the direction in which the hinge faces is the direction in which the tree will fall.

FIGURES 1 and 2 illustrate the hinge. The first is the view you would have if you were directly above the tree looking straight down

FIGURE 11. *Face cut closing as tree begins to fall*

FIGURE 12. *Undershooting the face cut*

FIGURE 13. *Overshooting the face cut*

FIGURE 14. *The ideal: the backcut is neither above nor below the undercut.*

FIGURE 15. *Avoid making backcuts too high or...*

FIGURE 16. *...too low.*

and equipped with X-ray vision; the second is a side view of the tree. (In these and other diagrams, straight lines indicate cuts in the tree, solid arrows indicate the direction in which the tree naturally will tend to fall, and dotted-line arrows indicate the direction in which we are trying to make the tree fall.)

The first cut that you make when felling a tree is the undercut (FIGURES 3 and 4). It should face exactly in the intended direction of fall and, in most cases, should go no more than a third of the way through the tree. This leaves room for wedges driven from the opposite side of the tree and reduces the chances that the tree will fall over backward. It is also important to make the undercut perpendicular to the trunk of the tree; angling the cut (FIGURE 5), since it leaves more wood on one side of the hinge than the other (FIGURE 6), may affect your accuracy in making the tree fall where you want it to.

You must also take care to make the undercut straight into the tree (FIGURE 7) so that your hinge will be a straight line. Otherwise, the tree may break off prematurely or swing to one side or the other. The best way to avoid making a crooked undercut (FIGURE 8) is to avoid making the cut with the tip of your saw's bar (FIGURE 9). A tree with a diameter greater than the length of your bar must be cut first on one side and then on the other, a technique that takes practice to master. If you are an able enough woods worker to cut trees that size, you probably don't need to read this chapter.

The second cut you will make is the face or notch cut. It starts a few inches above the undercut on the same side of the tree and descends at an angle until it meets the back edge of the undercut (FIGURE 10).

The face cut has little, if any, effect on the actual direction of the fall, but since it allows the tree to tip before the backcut meets the undercut, it is highly important (FIGURE 11). It takes practice to make the face cut meet the back edge of the undercut—even professionals miss from time to time. While you are learning, bear in mind that you would rather undershoot than overshoot (FIGURES 12 and 13), since you want to avoid having to make the undercut deeper. If you do undershoot, keep working at the cut until it's right; leaving it looking like FIGURE 12 (called "leaving a dutchman") could cause the tree to split when it begins to tip and the cut closes.

With one more cut—the backcut—you will have the tree on the ground. With luck. The backcut varies from tree to tree, as will be discussed later, but this much will almost always hold true: The backcut should begin on the side of the tree opposite the undercut and, ideally, be neither above nor below the undercut (FIGURE 14). In practice, it's okay if the backcut is a little high. Very high (FIGURE 15) or

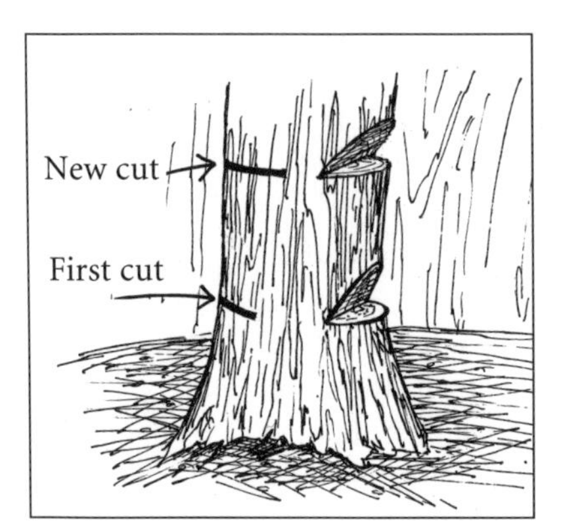

FIGURE 17.

FIGURE 18.

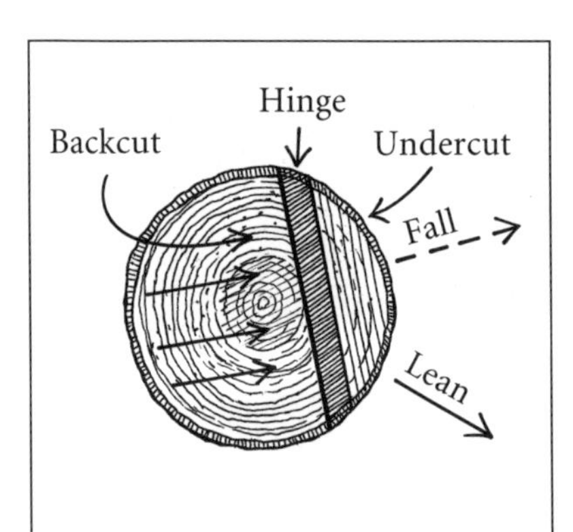

FIGURE 19.

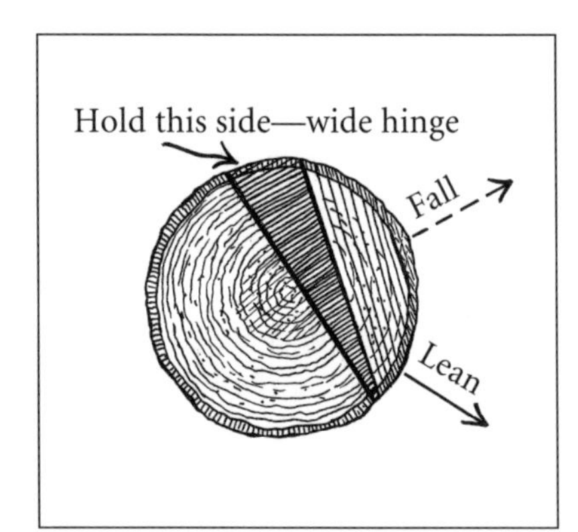

FIGURE 20.

low (FIGURE 16) backcuts reduce the hinge's effectiveness. If you do make a mistake with this cut, but catch it in time—before you've cut halfway to the undercut—you can start over (FIGURE 17).

Holding a Side

Now comes the technique that makes the backcut vary somewhat from tree to tree—holding a side. Often, when deciding where to fell a tree, you will see that if the tree falls in the direction in which it leans, it will end up in another tree or across a road or in some other undesirable spot. To avoid this, you have to alter the tree's course—often by only a few degrees (FIGURE 18). To do this you first make the undercut and face cut aim, as always, in the direction you want the tree to fall. Now, if the tree is a straight pole or has no large branches, you may be able to make your backcut parallel to the undercut (FIGURE 19).

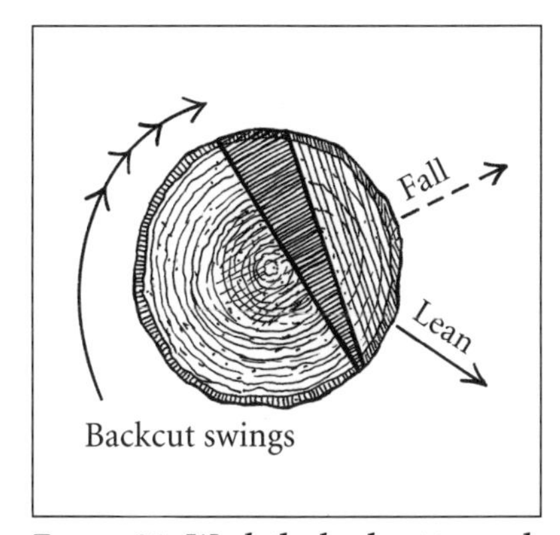

FIGURE 21. ***Work the backcut toward the side you wish to hold.***

FIGURE 22. ***Basic felling wedge application***

As is more often the case, however, the tree is not perfectly straight, has some large branches and leans enough to shake your confidence a little. In this case, you need more directional power than your undercut alone can provide, so you hold a side by leaving a little more of the hinge uncut on the side *toward* which you want the tree to fall (that is, the side *away from* heavy branches or a contrary lean). This is done by making the backcut at a slight angle to the undercut (FIGURE 20). It is important to emphasize that the wider side of the hinge will pull the tree in its direction, as in FIGURE 20, in apparent disregard for logic and physical order. There is a logical explanation for this bit of tree feller's alchemy: By leaving the side of the hinge that is away from the lean of the tree stronger, you are preventing the tree from breaking the hinge prematurely and falling where it leans instead of where you've faced it. Note, however, that if you cut all the way through the narrow side of the hinge, you will no longer have a hinge, and the tree may fall to the weak side.

In practice, you can rather easily hold a side by starting your backcut on the side where you wish to leave the narrow hinge and cutting it to within a half inch or an inch (depending upon the size of the tree) of the undercut. Then, working the backcut in the direction of the side you wish to hold, stop when there is an inch or two of wood (again, depending upon the size of the tree) between the backcut and the undercut (FIGURE 21)—unless, of course, the tree begins to fall before that point, in which case you may want to cut a little more to keep it from swinging too far or splitting, or you may want to get out of the way. It can also happen that you will cut your hinge exactly the way you want it, but that the tree will refuse to do the decent thing, instead remaining standing. Driving a wedge or two into the backcut will help it come to its senses.

Practice is the key to knowing how much to hold a side. The most common error probably is leaving too much hinge, resulting in a tree that doesn't want to fall at all, or a tree that swings too far. Practice on small trees when you have nothing (like your house or the power lines) to lose: You will develop a sense of how much you need to hold a side in order to swing a tree the desired distance.

Wedges

Crucial to the art of felling trees accurately is the wedge. While it is possible to work without them (beavers do), wedges give you a margin for error and allow you to handle trees that you might otherwise have to pass by.

In its simplest application, the wedge is placed in the backcut pointing in the direction that you intend the tree to fall (FIGURE 22). A larger tree might require two or more wedges (FIGURES 23 and 24). The wedge is usually placed in the backcut as soon as there is room for it. It is then tapped with hammer or axe until firmly seated. You then resume cutting, taking care to hit the wedge from time to time to keep the backcut from closing. When the backcut is deep enough, pull your saw out of the tree (you obviously can't do this if you're using a bow saw) and drive the wedge until the tree begins to fall. Sometimes the tree will fall when the cut is completed without further aid from the wedge. Sometimes, particularly with large trees, you may elect to leave your saw in the tree in case further cutting is called for after the wedge is driven home.

As a rule, I use wedges whenever I'm not quite certain that a tree will *not* fall backward, and also, in combination with other techniques already discussed, to make a difficult tree swing one way or the other.

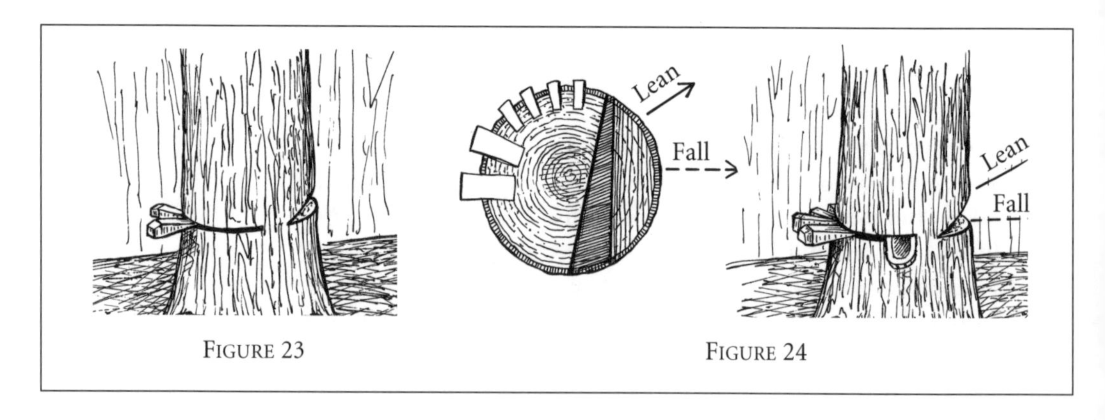

FIGURE 23

FIGURE 24

Until you are experienced enough to know your way around trees, I suggest that you use wedges as a matter of course. They can save your saw from getting bound in a tree that tips backward or maybe even save your life. Wedges will almost certainly help you drop trees where you want them, but they will not make a large tree that is leaning heavily fall opposite the direction in which it leans.

How to Do It

Now let's examine a number of different felling problems and their solutions.

We'll begin by assuming that you need ten cords of firewood and have obtained access to a woodlot. You have a chain saw, an axe and wedges, and you've selected the trees that you are going to fell. What now? First, if you've never felled a tree before, or only felled a few, limit yourself to trees that aren't very big to start with. By not very big, I mean 10 inches in diameter breast high (4½ feet from the ground) and smaller. A tree this size can still hurt you, but if you make a mistake, the mistake will be fairly easy to rectify.

Now you're standing in front of your first tree. Move close to it—close enough to touch it without stretching. Walk around it slowly, looking up. Does it lean one way? Which way? Which side has the heaviest branches? Check it for dead branches. Now check the base of the tree. Any obviously dead wood? Bear in mind that dead wood can't be used to hold a side.

Look around you. You have an idea where the tree wants to fall, but where do you want the tree to fall? Pick the spot with the fewest possibilities for hanging it up, particularly other trees with thick branches. Consider also your access to the tree once it's down: Felling is not an end in itself; the point is to get the wood. If your wishes and the tree's tendencies correspond, you're in business. If not, consider an alternative place to fell it. When you've reached a decision, remember to establish your escape routes and take one last look for dead trees and branches that might present a hazard to you. Now, commence.

The Straightforward Tree

This tree leans slightly in the direction you want it to fall. It's simple enough, but you'll still use a wedge in case you were wrong. You don't have to hold either side, but remember not to cut the sides all the way through. FIGURE 25 shows your approach.

The Tree that Doesn't Lean

This tree is similar to the last one in the way in which you

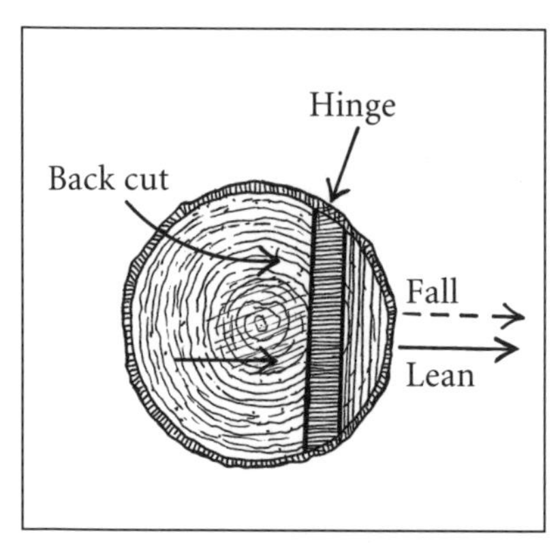

FIGURE 25. ***The straightforward tree***

FIGURE 26. ***Wedging a tree with no perceptible lean***

approach it. Be aware, though, that most trees do lean or at least have a tendency, because of their branches, to fall one way rather than another. So look again. Still pretty certain that it doesn't lean one way or another? Then go ahead and aim it where you want, but use two or more wedges, if there's room in the backcut; trees will fool you now and then. FIGURE 26 shows you what to do.

The Tree that Leans Heavily in the Right Direction

Hot dog! You've got it made, right? Wrong, maybe. With hardwoods in particular, a heavily leaning tree may split if no precautions are taken. This saves you the trouble of splitting the wood yourself, but it could also kill you if the split trunk jumps back or springs up. So take the precautions. There are two good ones, but I'll only describe one in detail because the second, though reliable, requires a bore cut—a dangerous practice only to be used by experienced sawyers.

What you do is this: Make your undercut and face cut as usual. Now, before you make the backcut, place your saw in the undercut and cut out one side, then the other, cutting toward the opposite side of the tree. How deep you make these cuts depends upon the size of the tree, but with a 10-inch tree, you would probably make each cut 1-2 inches deep. The reason that the side cuts work is that when a tree splits, it does so because the hinge is strong enough to momentarily withstand the enormous pressure exerted by the tree as it tips, so the pressure causes the tree to split vertically with the grain, at the point where the backcut stops. The side cuts simply reduce the strength of the hinge so that it will break before the tree can split. Now make your regular backcut, but cut just as fast as you can, and if you're

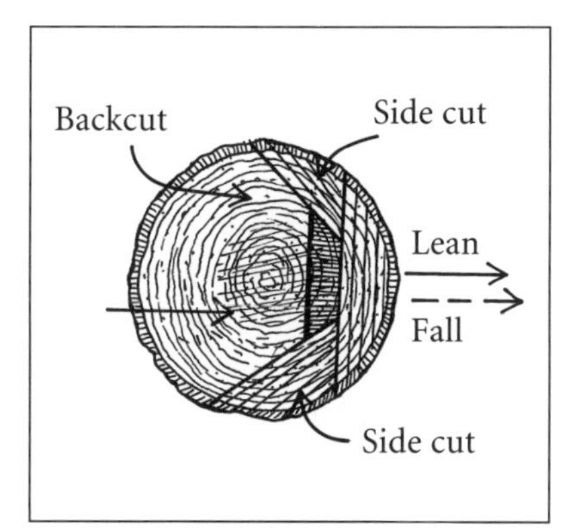

FIGURE 27. ***Side cuts keep a straight-grained tree from splitting as it falls.***

FIGURE 28. ***Wedging a tree that leans the wrong way***

nimble and not too frightened, keep cutting until the tree is really falling. That is, don't back off when it first starts to tip, as you might with another tree. Loggers, when they're cutting a big leaning hardwood of veneer quality, are so anxious not to ruin the tree that they'll keep cutting until the tree has bounced on the ground at least once. FIGURE 27 shows where to put the side cuts.

The second method of dealing with a heavily leaning tree is to make the undercut and face cut as usual, then bore into the side of the tree just behind the back of the undercut. Bore right through the tree, creating a narrow hinge, then cut your backcut from the hinge out: As you sever the last bit of wood in the back, the tree will fall, breaking the narrow hinge without splitting. As I mentioned, the bore cut is dangerous, as it carries a high potential for saw kick-back. Also, this method is impractical for small trees, since there isn't much room for an accurate bore cut, and it is very difficult to execute on a tree larger in diameter than your cutting bar's length, since you'd have to line up the bore cuts perfectly. Stick with the side cuts in this situation; like aspirin, they're safe and effective.

The Tree that Leans Slightly in the Wrong Direction

Your first step with this tree is to try to find a different place to drop it. Avoid conflict with the law of gravity whenever possible. If there really is no other spot and you have a couple or three good wedges, then go ahead as shown in FIGURE 28. The approach is quite simple, although you may need the third wedge, as shown in the diagram. Don't make your undercut very deep; in fact, make it a little shallower than usual unless the tree is very large. You need as much room as possible for the wedges. As always, be sure to leave a hinge. Otherwise, the tree will almost certainly fall over backward.

If you've done everything as described and the tree just stands there, you have four options: You can knock your wedges out and let the tree fall where it will. You can cut the backcut a little deeper—very carefully, so as not to cut off a side of the hinge. You can cut the undercut a little deeper (especially if the wedges preclude making the backcut deeper), using the bottom tip of the chain saw's bar. Or you can reach as high above the backcut as possible and cut an upside-down notch. Then cut a healthy, strong and long stick, put one end in the notch and push hard. Sometimes it works.

Another trick to try, if you have enough wedges, is driving two into the backcut piggyback—stacked one on top of the other. This can be ticklish, especially in cold weather, as they'll tend to pop out, so always have a couple of single wedges firmly in the cut when you try this. If all of the above fail, and you're beginning to feel that you're running out of luck and adrenalin, take the roll of brightly colored surveyor's ribbon that you have in your pocket and flag off a circle around the tree large enough in diameter so that nobody outside the circle could be hurt if the tree fell on its own. Then make plenty of "dangerous tree" signs with the marker and paper you have in your other pocket, and hang them from the ribbon. Then go get help from an experienced tree feller. Never leave a tree unmarked or unguarded in this situation: it is more likely—this must be a Murphy's Law?—to fall when you don't want it to than when you do, and there are enough people in the woods these days so that you can't assume that no one will walk under your tree in your absence.

Another trick for dropping a tree against its lean is to start the backcut first, drive the wedges in firmly, then make the undercut. This seems to have the effect of storing up the energy of the wedges, and will tip some trees with surpisingly heavy leans. It is also quite effective with trees too small in diameter for conventional wedging tactics. (With, say, an eight-inch-diameter tree, if you make the undercut first, the tree will likely tip back on your saw in the middle of your backcut before there is room to start a wedge.) The backcut-first method does take practice. Your backcut shouldn't go as far as halfway through the tree; if you feel the tree beginning to bind your saw, your backcut has gone far enough. When you first try this trick, you'll probably find that your backcut and undercut don't line up well, or that you make the backcut too deep and the tree begins to fall before you can make an accurate undercut, or that you cut part of the hinge off. This should tell you to practice with small trees that won't hurt anything by falling in the wrong direction. After you've gotten the hang of it, the backcut-first method will often come in handy, unless you never encounter trees that lean the wrong way, and we know what Mr. Murphy would say about that possibility.

The Tree that Must be Swung to One Side

This is the tree described in the section on holding one side. It's a tree that is not leaning in the right direction, but neither is it leaning in exactly the wrong direction.

When approaching this tree, don't expect miracles; a tree can often be swung 45 degrees, but more than that is very chancy.

For the actual techniques involved in swinging a tree, refer to Figures 20, 21 and 24 and the accompanying discussion. Two additional points to remember are:

1.) If the tree is hollow, dead or partially rotten on the side you want to hold, it probably will break off before it swings. Don't try it unless you have little or nothing to lose.

2.) If the tree has a pronounced lean and seems to be straight-grained, you risk splitting it, since you will have to leave a thicker hinge than you would usually leave on a heavily leaning tree. Be certain that you take care to avoid splitting it; you can use side cuts (refer to FIGURE 27) and hold a side, but that's a bit tricky, since you can't really see how much you're holding. For safety's sake, assume that any tree you cut is straight-grained enough that it might split if you make a mistake. If the lower part of the trunk is free of obvious twists and bends, it almost certainly is.

The Tree that Leans Heavily in the Wrong Direction

Forget it. Leave it uncut or find a better direction in which to fell this tree unless it's small enough to be pushed over. There are ways of handling a heavily leaning tree: Tree surgeons climb them and cut them in sections from the top down until they are de-weighted enough to be pulled in the right direction. Loggers cut them part way and push them over with their skidder or bulldozer.

There is one trick you can use on occasion, but it has drawbacks and should be tried only when you've gotten some experience under your belt. It goes like this: You locate another tree due for harvest which is at least as large as the leaner. This other tree must be one that can be felled so it hits the leaner hard enough to push it in the right direction. Approach the tree that leans and make the undercut and face cut. Then make the backcut as deep as you dare without having it close on your saw. (You could and probably should use wedges

FIGURE 29. *Using the peavey to rock a lodged tree loose*

here, but they may be hard to find in the tumultuous aftermath of this operation.) Now cut the other tree to fall so that it hits the leaning tree. They should both go over if your aim was good. If they don't, you're in a real pickle—a much worse and more dangerous dilemma than you had when you started—so you don't want to miss.

Problems

No matter how adept a woods worker you are, you will encounter your share of problems associated with cutting down trees. Trees will fall the wrong way and get hung up. Saws will get stuck. Tree felling is unpredictable work, and the most successful at it are the ones who know how to get themselves out of the inevitable jams.

Hanging Up a Tree

You've just felled a tree, and although you thought you were doing everything correctly, there it is—at rest in another tree. What do you do now?

The first thing to do is to cut the tree the rest of the way, if the hinge hasn't already broken. Sometimes this is enough to allow the tree to roll or slide backward (so get out of the way!) and dislodge itself. The next step, assuming that the first step failed, requires a peavey (sometimes erroneously called a cant-dog or cant-hook, a tool that is similar but doesn't have a point on the end). The peavey is a tool you should always have when cutting wood. It can be used as a log-roller or a pry-bar, and will get you out of many jams (FIGURE 29, inset).

Figure 30. ***Peavey as pry-bar***

Figure 31. ***Cutting a lodged tree: Don't get your saw stuck, and watch your toes!***

Take your peavey and hook the trunk of the tree. Try rocking the trunk back and forth (Figure 29). If this fails to dislodge it, try using the peavey as a pry-bar. Jam the point of the peavey between the butt of the tree and the ground and pry upward (Figure 30). This should allow the tree to slide backward and may be enough to free it from the tree that is holding it.

If you have a come-along, which is a lightweight, hand-powered winch, you can try hooking it to a nearby tree, then hitching the cable to your hung-up tree and cranking. A good come-along sometimes has enough power to dislodge a surprisingly large tree if you can set it up to pull in the right direction.

If all else fails, you'll have to cut the hung-up tree off, one chunk at a time, until you get it to the point where the peavey or come-along can finish the job. Start at a point not too far from the ground, perhaps two or three feet, and make a shallow cut from the top down.

Don't cut more than a couple of inches, or the tree will close the cut on your saw. Follow this with a cut from the bottom up until the chunk breaks off (FIGURE 31). Be very careful to keep your feet out from under the tree. Repeat this process until the tree comes down or until it becomes difficult to reach high enough to make further cuts.

If the tree does keep sliding down but doesn't fully dislodge, there may come a point when, with one more cut, it will be vertical or actually leaning away from the tree in which it is lodged. This is potentially dangerous, since the tree can now fall in a direction wholly of its own choosing—perhaps at you or a bystander. You can sometimes avoid this by, when you get to the critical point, making only one cut, and making it from the top down (FIGURE 32). The cut will close on your saw before you finish it, and when this happens, shut the saw off and apply a bit of pressure to the tree behind the cut, pulling your saw free at the same time. (Obviously, you shouldn't use this technique on a lodged tree that is too heavy for you to budge—to do so would be to add a stuck saw to your list of woes.) Now apply a lot of pressure at the same point or a point somewhat above the cut, pushing away from the tree in which yours is lodged. The cut you made will act as a hinge and allow your tree's butt to slide away from its vertical plane with the top.

If everything fails, there's not much else you can do, short of climbing the hung-up tree and cutting away the branches (very risky) or cutting down the tree in which it is lodged (also very risky). It's up to you to decide whether the tree, given its size and location, poses enough of a threat to the public to warrant hiring a vehicle with a winch to pull it down.

Cutting Trees or Branches that Are Under Pressure

You've just cut down a fair-sized tree and, while it isn't really hung up, it did hit a sapling and bend it double. Your tree is being held slightly off the ground by the sapling. What should you do?

Trees under pressure present a common hazard to anyone who works in the woods. When cut, they tend to jump rather unpredictably and very suddenly, and can cause injury or death. I've heard of people being killed by them, and I've been knocked cold by one myself. It isn't just trees under pressure that are dangerous; branches can also hurt or kill. They are bent and, consequently, put under pressure nearly every time a tree is felled (see Limbing, below).

Back to our case: You're alone in the woods (even though you absolutely shouldn't be) and there is your tree—almost on the ground but with that sapling bent under it, preventing you from limbing and

Figure 32. ***Pushing here will open the cut.***

bucking your tree. Your first move is to try to roll your tree away from the sapling with the peavey. If this fails, try pulling the sapling free, so long as it hasn't started to split; if it has, leave it alone, since it could break suddenly while you're tugging at it. Your third move is to get some expert help. Nine times out of ten, you could cut it yourself, but you won't recognize that tenth time, and I can't describe it, since it has so many variations. Watch and learn and be impressed, though not cowed, by the way a tree under pressure snaps and jumps when it is cut.

Now, suppose you've felled a tree and some branches on the underside are pinned to the ground, under pressure. This situation requires caution, but it is not usually so dangerous that you shouldn't cut the branches. The general rules are: Keep your feet clear, cut the branch close to the trunk and on the outside of its curve (Figure 33), since if you cut on the inside, the pressure will close the cut on your saw and leave you tearing your hair. Until you are comfortable with this kind of problem, cut slowly so that the branch will not release its pressure all at once. Come to think of it, pride goeth before a fall: It's not a bad idea to cut a branch under pressure slowly most of the time, regardless of how comfortable you are.

If the branch is large enough to hold the tree well off the ground, the situation becomes more dangerous, since the tree could roll when the branch is cut. Use your judgment. Always study the situation carefully, even if it seems simple and routine.

FIGURE 33. *Cut a branch under pressure on the outside of its curve.*

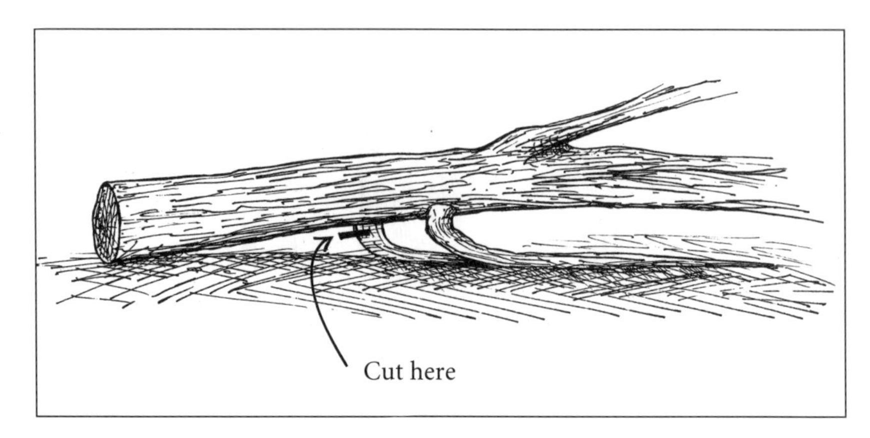

Freeing a Stuck Saw

This happens to everyone. Sooner or later your saw will become stuck while you are cutting, either because a tree leaned the wrong way or because you misjudged the pressure when you were limbing or bucking.

If it happens while felling a tree, it's probably because you weren't using wedges. While you are standing there with your $500 chain saw stuck in the backcut of a tree, you don't want to hear any I-told-you-so's, but you should have used a wedge. It may not be too late, even now; often, when a tree has tipped back on a saw, it's still possible to start a wedge and pound it in far enough to release the saw. If that doesn't work, and if the tree is too heavy to push, you'll have to cut it down with your axe or someone else's saw. To do this, make your cuts well above the stuck saw, and cut the tree so that it falls in the direction that it leans—no sense in getting two saws stuck. Try to disconnect your saw's power unit from the bar first, but if the chain is under too much pressure to release from the sprocket, say a few prayers.

If your saw gets stuck while bucking a tree that's on the ground, try to free it with your peavey. Jam the tip of the peavey under the tree as near to the cut as you can and, using the peavey as a pry-bar, lift. This should open the cut enough for you to extract the saw. If that doesn't work, try opening the cut with a wedge. If that doesn't work, cut through the tree with your axe or with another chain saw near where the saw is stuck. Don't get the second saw stuck by repeating the mistake you made the first time; if you bound your saw while cutting from the top down, release it by cutting from the bottom up (FIGURE 34) and vice-versa. Cut a little tentatively; pressure sometimes comes from more than one direction, and assuming the obvious to be true could leave you with a row of stuck chain saws: A good photo opportunity, perhaps, but otherwise unproductive.

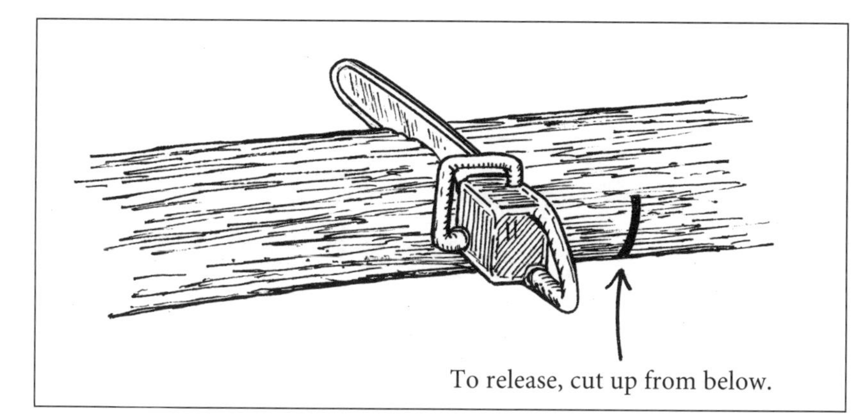

FIGURE 34.
Cut here to release stuck saw.

Summary

Felling trees is not simple and can be very dangerous. It's not bad to feel a bit intimidated by the job when you are beginning and, while felling will seem easier and will be more fun as you gain experience, it should always receive your full attention.

The situations covered here are fairly representative, but they are still hypothetical, and since each tree is different, I cannot stress enough the importance of starting out with an experienced woods worker to guide you.

LIMBING

Now that the tree is on the ground, it's time to cut off the branches—limb it. Larger branches are worthwhile firewood, of course, so before wading into the task, consider how the tree is going to get to your truck. If you are skidding it—dragging unbucked lengths by hand, with draft animals or with machinery—you'll want to leave the larger branches fairly long or, if terrain and equipment allow, perhaps leave them attached to the trunk.

If, on the other hand, the tree has been obliging enough to fall close to your truck, you'll probably be bucking the wood to stove or fireplace length on the spot. These two situations favor different limbing strategies: Trees that will be skidded are generally best limbed from the butt to the top, because in this way the sawyer can better avoid both working in brush and long reaches with his or her saw. Trees that are to be bucked in place are usually better limbed from the top down, because the larger branches and, usually, part of the trunk can be cut while they are off the ground—a blessing to both saw chain and sawyer's back. Naturally, these rules do not apply uniformly: Variations in trees and terrain and the lie of the felled tree can dictate changes in limbing strategy, so familiarize yourself with both.

FIGURE 35.

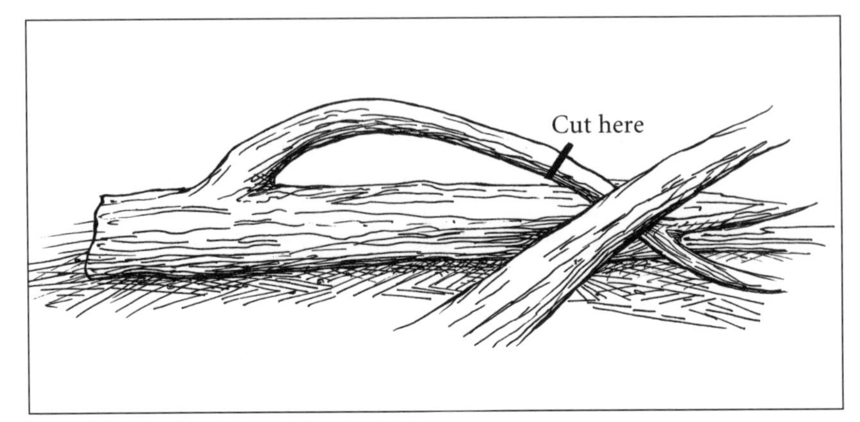

FIGURE 36.

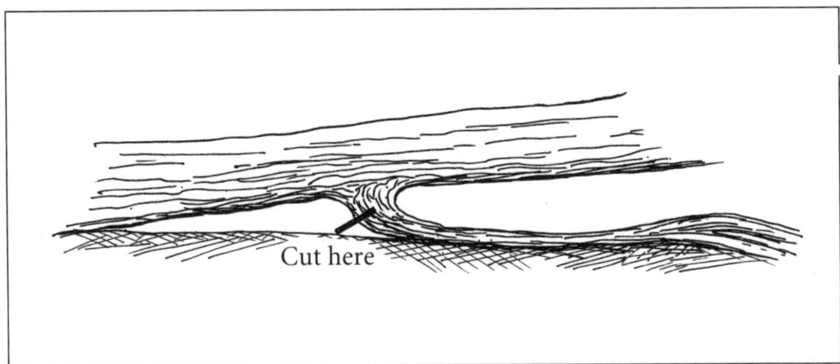

Before examining the nuts and bolts of the limbing techniques, here are some general safety rules:

1.) Before cutting any branches, study them all; which ones are under pressure, which are holding the tree up, which are preventing it from rolling?

2.) Whenever possible, cut branches on the side of the tree away from you, thus keeping the trunk between you and your saw's chain.

3.) Keep your working space clear; toss brush out of your way as you go. Tripping is one thing; tripping with a chain saw in your hands is quite another.

4.) Avoid cutting with the tip of the bar; saw kick-backs occur frequently during limbing; so cut one branch at a time, and be certain that you know where the saw's bar tip is and that it is free of obstructions.

Limbing from Butt to Top

The hazards specific to this limbing technique are that the sawyer is usually within reach of the tree should it roll, and the branches will often be apt to snap and spring when cut due to pressure exerted by the tree's weight or by contact with other trees. Before you start cutting, identify the potential troublemakers and devise ways to deal with them (see Cutting Trees or Branches that Are Under Pressure, above). If, for example, a large branch has been bent back by another tree and obviously is under pressure, I suggest carefully cutting and removing from the work area every branch that is positively not under pressure or potentially helpful (e.g., It might keep the tree from rolling toward you). Now that the situation is less cluttered, consider: Where can you cut the branch to relieve the pressure without putting yourself in jeopardy? Will cutting it cause the tree to roll? Which way?

Scary they can surely be, but as you gain experience, you will find that branches and trees under pressure will usually yield nicely to a patient, considered approach. Never take foolish risks, though: If you don't know what will happen when you cut wood under pressure, don't cut it! FIGURES 35-39 show branches under different types of pressure and the correct positioning of the cut in each case. Please remember, however, that trees pose a nearly infinite variety of hazards and that diagrams can neither identify them all nor do them justice. Assisting and learning from an experienced woods worker is your best bet, particularly if that training is blended with a bit of valorous discretion. Seasoned pros—usually confident and in a big hurry—have been killed by branches that released their pressure suddenly and unexpectedly. When this happens, too often the first thing that hits you is your chain saw, followed in 1/64th of a heartbeat by the butt of the branch that you were cutting.

Limbing isn't always a battle with the grim reaper, but I hope I've impressed you with its hazards: It's probably a tendency for neophytes to be so relieved at having safely felled the tree that leads them to take the limbing for granted.

Now for the job at hand. We'll assume that the tree is straightforward, with only the branches under it being under pressure. We'll also assume that you are right-handed; lefties should start on the other side of the tree.

Step 1: Standing on the left side of the tree, reach over the trunk and cut any branches on the opposite side within easy reach. (I emphasize "easy" for two reasons: Operating a chain saw with your arms extended is tiring, and it can also be dangerous, as you will have less control in the event of a kick-back.) As the contours of the tree permit, let the trunk support your saw's weight (FIGURE 40).

Step 2: Stepping back slightly from the tree, bring your saw up and, still resting some of its weight on the trunk, cut any top branches within easy reach (FIGURE 41). You may find that you are more comfortable using the top of the bar for this cut—to do so saves

FIGURE 37.

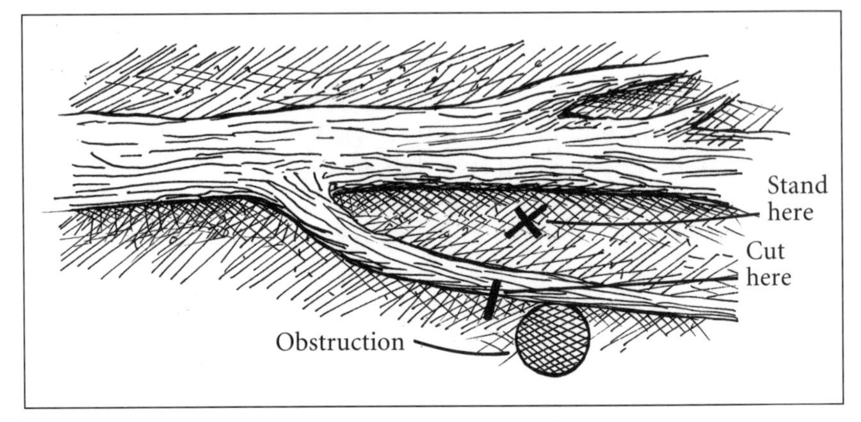

FIGURE 38.

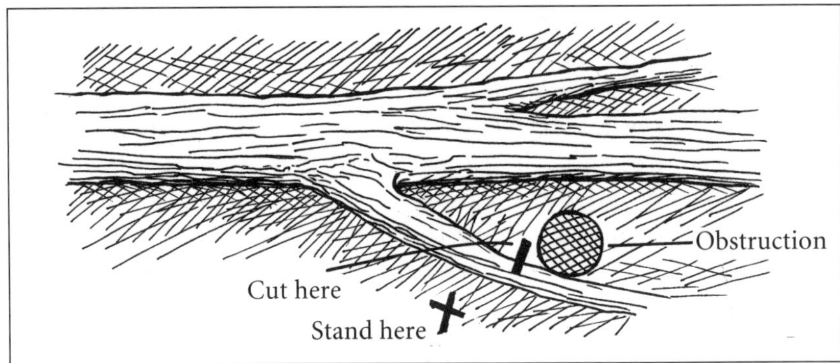

FIGURE 39.

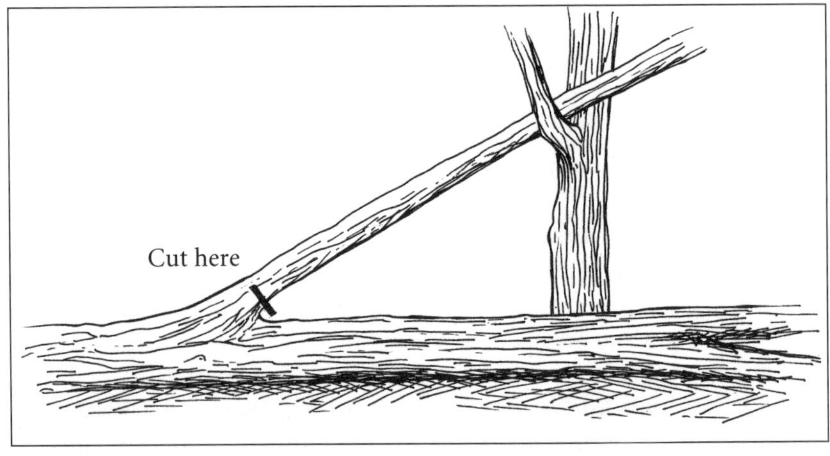

turning the saw and tends to throw the sawdust away from you. Just be careful to keep the upper tip of the bar in the clear: This is a bad time to deal with a kick-back.

Step 3: Step back a bit more and cut the branches on your side of the tree (FIGURE 42).

Step 4: Finally, reach under the tree and cut any branches you

FIGURE 40. *Let the tree support the saw's weight.*

FIGURE 41. *Using the top of the bar to cut a top branch*

can without your saw's hitting the ground. These cuts will be made, usually, on the outside of the branch's curve (FIGURE 43). Be careful, because the branch is under pressure and could jump or cause the tree to settle or roll or both, possibly pinning your saw.

And that's it. Keep repeating the sequence until you reach the top, in the order I've described as much as possible, since, besides being fun and rhythmic, proceeding that way tends to allow cut branches to fall out of your way, minimizing the number of times you have to stop cutting and handle brush. Small trees can't reasonably be used

FIGURE 42. *Step back and cut branches on your side of the trunk.*

FIGURE 43. *Cut carefully when releasing pressure.*

to support your saw in Steps 1 and 2, but should otherwise be treated the same as larger trees.

Limbing from Top to Butt

When done correctly, this limbing method offers a considerable built-in safety feature: The sawyer is almost always standing where the tree cannot hit him/her if it rolls. The disadvantages of the technique include that, when applied to larger trees, it often requires the sawyer to cut above shoulder level—a no-no for all but the most skilled and

FIGURE 44.
A cut here will release the pinned branch, causing it to spring up suddenly.

fit—and that much of the tree's brush falls into the work area and must be periodically cleared to assure safe footing. The dangers posed by branches under pressure are still present with the top-down method, but usually involve the small end of a pinned branch suddenly springing when it is released (FIGURE 44). This is certainly dangerous, but the direction in which the branch will spring is often more predictable when it is released from the top than when it is cut at its base, so sizing up the cuts before you make them can remove much of the thrill and danger of the unknown. Assuming that we're dealing with a tree that poses no extraordinary problems, here's the step-by-step approach:

Step 1: Standing near the top of the tree, on the right side as you look from the top down (left side for left-handers) and turned slightly toward the butt of the tree, cut the brush off any branches that you can comfortably reach and that are not under pressure.

Step 2: Buck firewood lengths, if any, from these branches back to the trunk.

Step 3: Clear the brush you've cut, and repeat the process on the other side of the tree.

Step 4: Cut branches on the top of the tree, if they're within safe reach, bucking firewood lengths as you go. It is often possible to bring a large branch within reach by cutting it partly through where it meets the trunk, thus letting it down.

Step 5: Cut firewood lengths from the trunk and repeat the entire sequence until you reach the first significant branch that is under pressure—very likely a bottom branch (FIGURE 45).

Step 6: Standing clear of the trunk, work the bottom branch back until the pressure is released, usually by making a series of undercuts

FIGURE 45.
Buck firewood lengths back to branch under pressure.

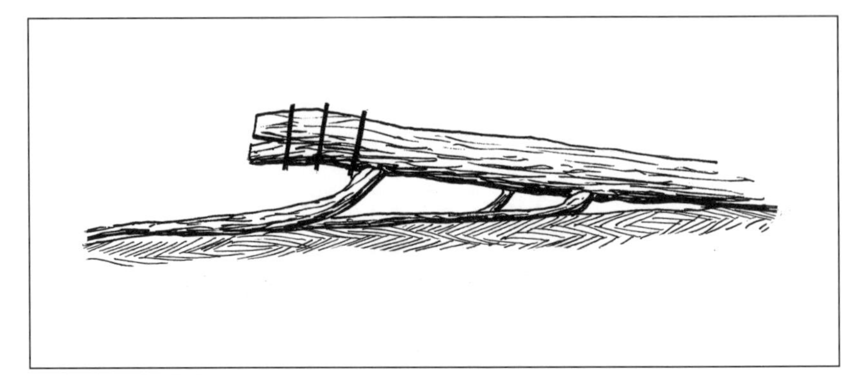

FIGURE 46.
Release pressure with a series of cuts.

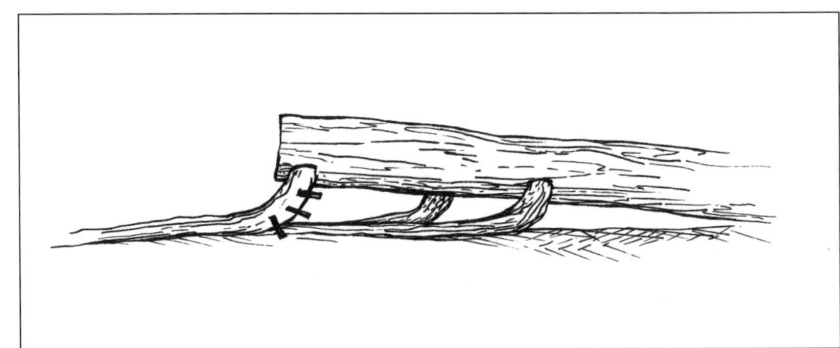

FIGURE 47.
Woodlot road system

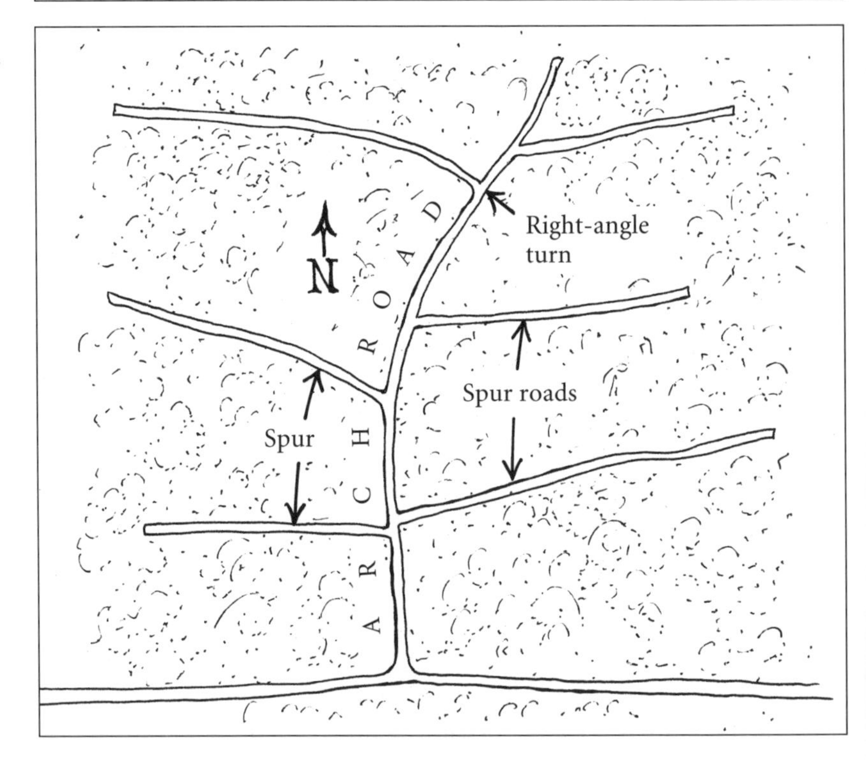

(FIGURE 46). If the branch is so close to the ground for most of its length that you can't undercut it, carefully cut it at the trunk on the outside of the curve, after first ascertaining that doing so won't cause the tree to roll at you. Repeat the steps until the tree is limbed, bucked and ready to split.

SKIDDING

Skidding, or yarding, is the process of moving wood—usually but not always in long lengths—from the woods to a place where it can be loaded on a truck. In a commercial logging operation, skidding is the most expensive part of the job, largely because of the equipment required. It is also the most critical limiting factor—assuming that there is a good market for the wood—since the feasibility of skidding (How far from the stump to the truck? Is the terrain passable?) will determine whether it's worth cutting a particular woodlot. For a do-it-yourself fuel wood operation, the problems are the same, but on a far smaller scale.

Skidding Techniques

It's probably harder to describe how to skid trees than it is to describe how to fell them: Not only is every tree different, but so is every skid trail, and for that matter, every day. But in order to give the uninitiated an idea of what skidding is and, in the process, impart a few techniques from my limited store, I'll describe two different hitches that might have been made with a skidder, a hitch being one trip out of the woods with a load of timber. (The load itself is also called a hitch, but this is not really confusing in practice.) Both hitches are perfectly typical—the first is uneventful and the second a bit fraught with hair-pulling little problems. Of course, you probably won't be using a skidder, but the principles are the same for tractors, winch-equipped trucks and, to some degree, draft animals.

We've set up the road system in this woodlot quite logically: Our main skidding road—the arch road—is fairly straight and goes from the landing (a clear space where the truck has access to the wood) to the back boundary of the property, roughly bisecting the lot. We've cut a few spurs, as needed, to give ourselves access to major concentrations of timber, and are working from the back end of the property out in order to avoid unnecessary wallowing in our own brush (FIGURE 47).

On this particular hitch, I'm going to the northeast corner of the lot. I drive almost to the end of the arch road and take the last spur to the right, because I know that this is where the chopper, who is doing the felling and limbing, is working. As I drive down this road, the chopper steps out from the side and indicates that the hitch is just off the road, so I turn around and back into the spot he's pointing out. There, in a nearly perfect semicircle, are the butts of four good-sized hardwood trees which have been limbed and topped. The chopper and I attach choker chains, or chains with a choke hook on one end that are wrapped around logs for skidding. The other end of the chain is secured to the skidder's winch cable by a grab hook that is on a slide on the cable. We chain all four trees, and I pull the winch cable out and hook the chains to it. Then I winch the trees up until they're against the back of the skidder, their butts well off the ground, and away I go—back the way I came, to the landing, where I'll unhook the trees, buck them to length and push them out of the way and/or sort them by product type. Then it's back to the woods for the next hitch.

Another day, another hitch. This time we're in the northwest corner of the woodlot. I make the turn, noting that it's almost a 90-degree angle and none too wide, and hoping that that won't cause problems. I drive down the spur until I see the chopper, who comes up alongside and says that the hitch is off of the road a bit. I back in off the spur road as far as I can—quite far, actually, because the chopper has brushed out the access, but when I get to a wall of closely grouped trees, I realize I've reached the end of the line. I can see the butts of a couple of downed trees through the woods, maybe 75 feet away, so I grab the chokers and the end of the winch cable and head in that direction.

These trees aren't arranged in a perfect semicircle; in fact, they're scattered and separated by other trees. After looking the problem over for a moment, I know that I won't be able to simply hitch up all four trees, winch them in at once, and go; they'd never make it through the standing trees. If I play my cards right, however, I won't have to winch each one separately.

First I go to tree Number 1. Its butt is directly against an obstinate-looking rock outcropping, so I attach the choker chain in such a way that the hook, instead of being on top of the log, is nearly on the ground on the side of the log away from the skidder. This is a "rolling hitch," as becomes apparent when I hook the choker to the winch cable and winch the tree in: The strain on the hook causes it to rotate and roll the log free of the rock. I winch this tree until it's clear of the trees between it and the other three trees that will make up the hitch, then unhook it and consider my next move.

Tree Number 2 cannot be pulled straight toward the skidder without having it butt up against a large standing tree. Getting it around that obstacle would require setting up a snatch block or doing a good deal of peavey work, so instead I hook trees 4 and 3 and, finally, 2.

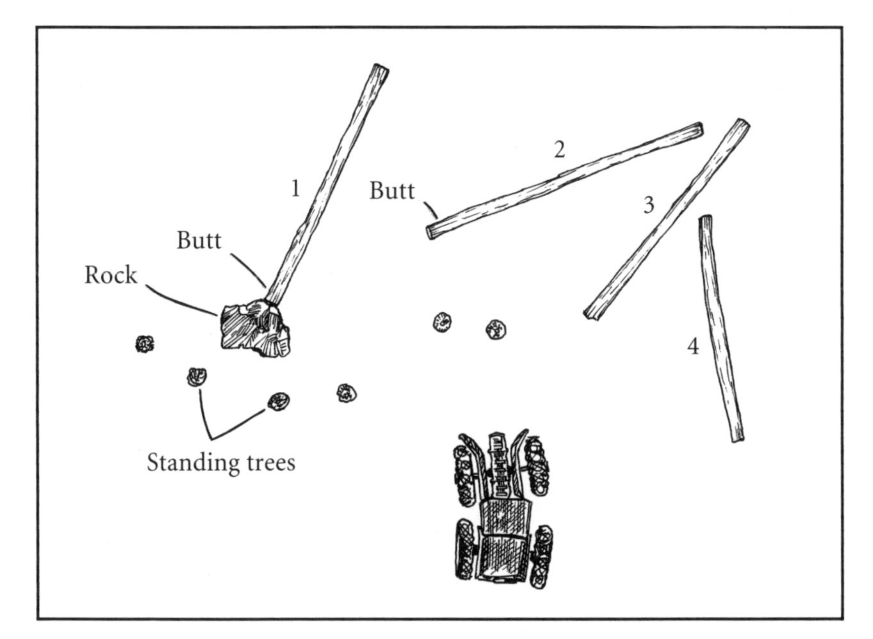

FIGURE 48. ***Pulling around obstacles***

When I power the winch, the weight of 4 and 3, acting as a de facto snatch block, draws tree 2 over to 3, clear of the obstruction. It's then a simple matter to rehitch tree 1 and pull all four up to the skidder. So far, so good. Sometimes, of course, the weight of the other downed trees is insufficient to allow the winch to pull the problem tree clear, or there might be no downed trees where you need them. In such a situation, you could set up a snatch block (a pulley) on a standing tree and run the cable through it. This will change the direction of the pull long enough to get your tree clear of obstructions, at which point you can take the winch cable out of the snatch block and winch straight to the skidder (FIGURE 48).

Back to the skidder: As I drive up out of the woods toward the spur road, the combination of the enormous weight of the load, the uphill pull and the soft ground makes the tires begin to spin. I'm losing headway, and it's apparent that I'll have to try something different if I intend to bring this whole hitch to the landing at once. So I release the winch—put it in neutral—and drive up to a level spot near the road, the cable automatically paying out as I go. At this point I stop and winch the trees up to the skidder. Here, past the steep going, I'm able to drive away with the hitch.

The next problem arises, as I had feared, at the intersection of the arch road and the northwest spur. We hadn't made the corner wide or gradual enough, and now, as I try to go around it with four 75-foot-long trees in tow, the tops catch in the standing trees on the far corner. I try releasing them and winching them up from a different angle, but to no avail. I don't have a chain saw with me, so I can't cut the tops off of my hitch even if I wanted to. What I do instead is unhitch one of the trees—the one that is hanging up the most tenaciously—and leave it. The three remaining trees in the hitch now pull free, and the one I left will serve as a bumper; on subsequent hitches, the tops will slide off of the bumper instead of getting caught among the standing trees. This doesn't always work, either. But what does? Sometimes you'll have to get a chain saw. (Watch out for pressure when you make this kind of cut!) Sometimes you may keep pulling until something breaks. This is skidding.

With draft animals, of course, there are no winches, so you'll be backing up to every tree you skid, though you can hook several chains together to pull a log within reach. You'll also seldom if ever be pulling tree-length hitches, so a lot of problems—such as tight corners—won't plague you often. You will, however, be obliged to spend quite a lot of time clearing trails for your critters, and they won't move as much wood as a machine, but this, too, is skidding. (More on animals below.)

Skidding Roads and Waterbars

Waterbars deserve mention at this point. This is not a skidding technique so much as it is a road- and soil-maintenance technique, but taking proper care of your skid roads is as much a part of the operation as moving the wood.

Skid roads, particularly well-used ones, tend to lose their vegetative cover. If they are anything other than perfectly flat, they are vulnerable to damaging erosion. I've seen old hillside logging roads that look like stream beds, and that's a sad sight. The solution—waterbars—is easy. You simply dig shallow trenches diagonally across the road on any steep sections. The trenches divert water to the side where vegetation and uncompacted soil can handle it. Waterbars should always be installed upon completion of a job, but sometimes you'll need to build and maintain them while the road is still in use. They're a bit of a nuisance to skid over, but nowhere near as much of a nuisance as a road surface that migrated to the foot of the hill in yesterday's rainstorm. Talk with a forester about construction and placement of waterbars.

Skidding Rules

"We've found that the best solution to the problem of skidding firewood is a good road system." So says Gary Salmon, a Vermont state forester with many years of experience in administering programs that make firewood available to the public. What he means, of course, is that being able to drive your truck to the tree is far more practical than dragging the tree to the truck. And he's absolutely right: Any piece of equipment that is capable of fast, efficient skidding under a variety of conditions is far too expensive to be of practical use to a homeowner trying to supply his own firewood needs. Having said that, we must acknowledge that good woodland road systems aren't available to all woodcutters, and that even where they are available, they don't eliminate the need to move wood to the truck, they just reduce it. Also, there is good wood to be found in places where neither roads nor trucks would wisely venture, so it's worth examining the ways of moving the wood to the truck. Some involve bulling and jamming and backache, some involve considerable cash outlay, and some involve a tinkerer's soul and a high tolerance level for frustration.

Here's a list of a few good rules to follow:

1.) The shortest distance between two points is a straight line: Avoid skidding around corners whenever possible.

2.) Stubs, brush and rocks in the skidway never look significant until you try to pull a log over them: Remove obstructions in advance to avoid frustration and unnecessary wear and tear on equipment and muscles.

3.) Getting the butts of logs off of the ground, whether by chaining them to a stoneboat or winching them up tight to the skidder, makes it much easier to drag them.

4.) Following our forester friend's advice, where it's feasible and responsible to do so, installing a good road or road system to shorten the skidding distance will pay handsomely: Long skidding distances drastically reduce productivity.

5.) Skidding systems that require extensive setup time, such as cable yarders and chutes, require a substantial concentration of harvest-ready wood to be practical. For that matter, it makes no sense to cut roads unless they provide access—now and in the future—to a significant quantity of wood.

6.) Our last and most important rule is: Plan your work carefully. Irresponsible, thoughtless skidding—even with draft animals—can do significant damage to remaining trees, soil and water crossings. It is ethical to use wood only if those who cut it take the measures necessary to keep it a renewable resource.

Hand Tools for Skidding

Your hands themselves are tools, of course, and with them (and your back and legs) you can move a surprising amount of wood a surprising distance from stump to truck. How much wood you can move and how far you can move it depend upon your level of fitness and determination and upon the terrain involved. How far you *should* move wood by hand is a question you'll have to answer for yourself. For me, 100-150 feet over reasonably good terrain is about the limit; beyond that it takes too much time to fill the truck and ceases to be enough fun to justify as a hobby.

I arrived at this conclusion the hard way: I was given the opportunity to cut a three-cord fuel wood lot in the state forest. The purpose of the cut was to create a wildlife opening, and the trees to be cut were eight large sugar maples, all within 200 feet of the road, but all down a slope of 45 degrees or more. Sugar maple for $6 a cord was too good for my greedy heart to pass up, so naturally I cut it and hauled it up that slope. Naturally I did it during an unseasonable heat wave. And naturally, it was blackfly season. My Newfoundland dog, who came along to keep an eye on me and the truck, got bitten so badly that the ASPCA would doubtless have confiscated him, had they happened along, and rightly so. I fared no better. In retrospect, it was probably for the best that I cut that lot: I learned an important lesson, and had I not, I might have made a fool of myself for some inferior species of firewood.

Over the years, I've found a good deal of wood within a reasonable (both for distance and terrain) haul of truck access, and I've also found a couple of hand tools that make the transition from lazy human to beast of burden a little easier than it would be bare-handed.

The pulp hook (FIGURE 49, bottom illustration) is about as simple as tools get: No batteries, no motors, no moving parts at all. As the picture shows, it's a wooden handle with a metal hook attached. Used for years by pulp cutters to pile four-foot wood and, in the days before hydraulic loaders made life a bit easier, to load it onto trucks, the pulp hook gives the user no exponential increase in leverage. It does provide a secure grip, welcome always and critical when wet or icy wood is being handled. I often use a pulp hook to double the load I carry: I'll put one stick on my shoulder and drag the other behind with the hook.

I have a running disagreement with a friend who hates pulp hooks. He maintains that they are dangerous, and I do have two scars (forehead and knee) that would seem to support his theory. I got both scars when I was young, however, and didn't know enough to avoid breaking two of the three cardinal rules of pulp hooking: 1.) Don't use a sharp-pointed hook (a mushroom point is best because it stays where you stick it); and 2.) Keep a safe distance from other pulp hook operators. The third rule? Don't leave your hook where you can run over it with your brand new radial tires.

To use a pulp hook, swing it sharply enough so that the tip seats securely in the stick of wood you wish to carry, drag or throw. Practice makes less imperfect: If you stick the wood too enthusiastically, you'll use a lot of energy getting the hook out, particularly if you were throwing that particular piece and had to follow it at high speed because the hook wouldn't release. Conversely, a pallid swing may allow the log to drop on your foot in transit. You'll notice, also, that different species of wood require different amounts of force to seat the hook; the swing necessary to get a firm grip on a stick of beech would probably wed the hook to a stick of pine for the duration. Where to place the hook in the wood is partly a matter of function and partly a matter of individual preference.

One way to use a pulp hook is to stick it in one end of the log, stand the log partly upright, grab it near the other end with your free hand, and carry it away. If you don't have far to go, you can throw the stick, using the hook planted lightly in the end, for leverage. If you prefer to drag wood, plant the hook in the top of the log—near the end but not in the end—with the hook pointing in the direction of travel so that it doesn't unseat itself.

The other useful hand tool, the log carrier (FIGURE 49), is a stout hardwood handle, approximately four feet long, with a pair of hinged tongs suspended on a swivel from the middle. This piece of equipment requires you to find a helper, but so much the better if you can share the misery. The helper should, ideally, be nearly your

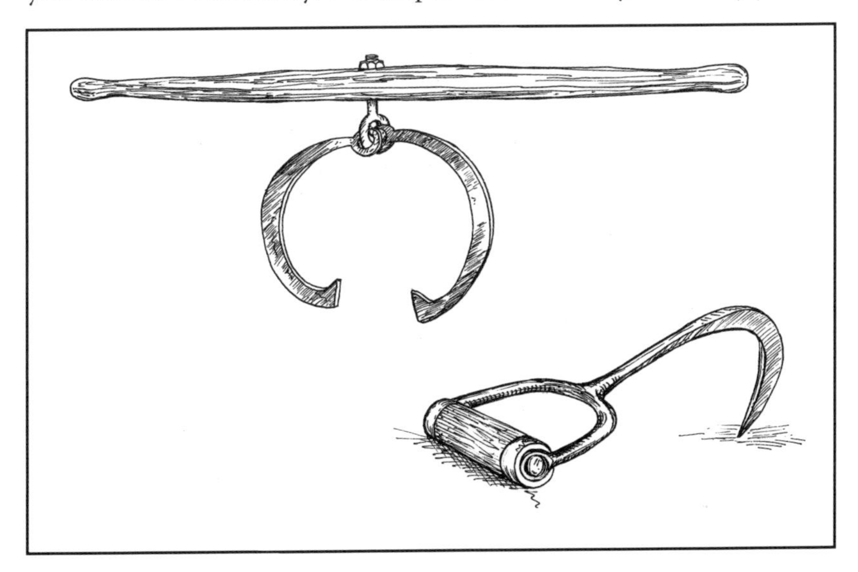

FIGURE 49. ***Pulp hook (bottom) and log carrier***

height and strength, because otherwise it can be very difficult to coordinate the effort. If you've ever been to the horse pulling contests at a county fair, you'll know that mismatched teams rarely take the blue ribbon.

Operation of the log carrier is simplicity itself: Hook the tongs in the large end of the log, tap them to make certain that they're seated, and off you go. Two reasonably strong people can move big pieces of wood this way, and the spirit of competition will probably make them keep at it long after good sense would have prevailed had they been working singly.

Chain Saw Winches

For around $500, you can get a well-made, lightweight winch which, when attached to your chain saw's powerhead (you have to remove the bar and chain to do this) and strapped to a tree or stump, can reach 150-300 feet (depending upon the type of cable used) into the woods and pull out several hundred pounds of wood at a time. I've seen—and been impressed by—one model, the Lewis winch. The tool's best application is yarding wood a distance of one length or less of its cable: Resetting the winch and hitching every stick twice or more would certainly dim this tool's luster for me. I also suspect that, if you got a chain saw winch, you would very soon tire of removing your bar and chain and would buy another saw to use as a semipermanent powerhead. A chain saw winch will also dislodge hung-up trees and stuck vehicles, among other things, so it looks to be a handy tool if you have the work for it. Every silver lining does have a black cloud, however: It's an unusual woodlot that consistently allows you to reach 150 feet straight in with a cable and pull anything out without encountering obstructions. Since a chain saw winch works most efficiently from a fixed position, you need to plan and clear skidding corridors carefully. Constantly resetting the winch could make a pulp hook or log carrier look like a time- (if not labor-) saving device.

Truck-Mounted Winches

By far the most common type of truck winch is the electric front-mounted type. Uncommon, but not impossible to find, are military surplus tactical vehicles equipped with heavy-duty hydraulic winches, but as these trucks are usually quite large and not very practical as over-the-road vehicles, their main appeal should be to people who own substantial woodlots and are willing to put in comprehensive road systems. Under these limited circumstances, a surplus all-wheel-drive truck with a hydraulic winch can function quite nicely as a skidder, as long as you remember that it isn't really a skidder; all-wheel- drive or no, these vehicles venture off the road at their peril. Also, since their winches are generally front-mounted, you'll want to unhitch each log after it's winched to the road, turn the truck around and back up to it, and chain it to the back unless you don't mind driving in reverse to the wood's destination. This certainly can be done, but it gets boring.

Another way of using the surplus truck and winch, which makes a good deal of sense as long as you can drive the same vehicle to the door of your woodshed, would be to winch the trees to the road, cut them on the spot and load the wood on the bed of the truck. In this way you eliminate some of the problems inherent in dragging tree- or log-length wood behind a vehicle that wasn't designed for the purpose. For example, the above-mentioned turning around and rehitching business becomes unnecessary. You also avoid a more vexing problem in skidding logs with a truck: lost logs. Since you can't get the butts of logs very far off the ground when you chain them to the back of a truck, the logs have a tendency to come unhitched on their own. Arriving at your destination only to find that half your cargo chose to jump ship early is tough on the blood pressure. I reiterate, however, that as well as an old surplus truck with a winch can work, without a good road system the vehicle will probably take up more space than the wood it skids for you.

Everything negative that we've said about surplus trucks and hydraulic winches is true in spades of standard 4X4 pickups with electric winches. The trucks, tough though they may be, simply weren't designed or built for log skidding, and neither are their electric winches, which, by the way, are classified as "vehicle recovery winches." It's one thing to drive through a mud hole or pull a stuck car out of a ditch; it's quite another to haul wood over rough terrain for hours on end. Wood becomes very costly when extensive truck repairs are figured into the price.

Certainly, if you have a pickup truck with an electric winch, you may find from time to time that it comes in handy when you're cutting firewood, but I doubt that you will have much success basing an entire wood-gathering operation on this sort of equipment.

Skidding with Tractors

Farm and utility tractors, particularly four-wheel-drive models, are often used to skid firewood and logs. When equipped with a rear-mounted, power-take-off-driven winch mounted to the tractor's three-point hitch (allowing you to raise the ends of logs off the ground), and proper roll-over protection, a tractor can do a pretty fair imitation of a real skidder. So equipped, of course, the tractor's price will also be imitative of a skidder's, which is to say that buying a tractor equipped for skidding is impractical unless you intend to cut commercially. Given the price tag, winch-equipped tractors are really only appropriate for people with access to large woodlots and the time and energy required for serious harvesting.

If, however, you already own a tractor and own, or have access to a woodlot, you might consider spending the $1,500-$4,000 (depending upon the size of the tractor) for a winch. There are a number of different brands, and you'd be well advised to shop around and take into account the availability of service before choosing, since it will take a few years for the winch to amortize itself. But make no mistake; even mounted on a small (17-horsepower) four-wheel-drive tractor, a winch will allow you to skid significant amounts of wood and will give you practical access to parts of your woodlot you'd otherwise bypass.

Draft Animals

Many books have been written on this subject, and it is not my intention to write another, but no discussion of wood skidding is complete without a mention of draft animals. When I say draft animals, I refer to horses and oxen. Pigs are certainly strong enough for the work, but are probably too smart to be lured into harness.

I very much like working with draft animals: I've logged with horses (as the sawyer, however, never as the teamster, so my enjoyment may have been augmented by a bit of distance) and been surprised over and over again at what well-trained animals can do. I am currently helping my wife train an ox, and this experience is giving me a large measure of respect for the time, patience and steadiness demanded of anyone who trains and works draft animals.

Animals have a lot to recommend them, if your circumstances are appropriate: They can skid relatively large amounts of wood, do relatively little damage, don't cost as much as machines, encourage agricultural use of land, don't use gas and are fun and rewarding to work with, if you are of the right frame of mind.

The other side of the coin is that draft animals require considerable facilities (barns, pasture, fence, etc.), are not always easy to transport if your woodlot is farther than walking distance from home, eat whether they're working or not, need daily attention and are, for all the time and work they require, impractical for skidding distances of more than a few hundred yards under normal conditions.

If you enjoy working with animals and your property includes ample pasture and wooded land, it may be that draft animals would be the best possible addition to your wood-gathering program—"best" in this case meaning most rewarding, not necessarily most efficient.

You would certainly do well, if you haven't owned or worked draft animals, to make the acquaintance of someone who does—someone willing to help you acquire a bit of firsthand experience with the trials and tribulations involved in the undertaking. After building a barn and fences and buying 5,000 pounds of Belgian horses and their attendant tack would be a hard time indeed to discover that you didn't like working the animals.

From watching experienced teamsters working with animals in the woods—and from my role in the work—I've drawn the following conclusions: 1.) It takes well-trained animals to work well in the woods, but of equal importance is the training of the person working them and the relationship between the teamster and team; you have to know how to get enough out of your horses or oxen without asking too much. Many draft animals will try to do the impossible if asked, but repeated failure can ruin a creature's spirit. 2.) The person doing the cutting has a good deal more to worry about than just felling and limbing: Draft animals need clear trails, room to turn and back up to the log, and reasonable terrain. The question for the sawyer is not just where to fell the tree so that it will hit the ground; it is where to fell the tree so that Dobbin and Charlie can get at it. This may involve more creative felling and more painstaking trail clearing than other skidding methods require, but woods work always involves trade-offs: If you had a big rubber-tired skidder, you could drop the trees any which way and pay little attention to trail clearing, but your woodlot would look like hell when you were done.

Exotic Skidding Systems

"Exotic" may seem a strange adjective to apply to the following systems, composed as they are of such mundane materials as rope, pulleys and pvc pipe, but the ingenuity which led to their creation—and which, in some measure, is required for their operation—makes the adjective appropriate, particularly when these systems are viewed in comparison to horses or tractors. Note that all three systems—ground cable, suspended cable and chutes—are fixed in place, which is to say that the operator does not drive from tree to tree but must break the rig down and reset it in a new location once the cutting in a particular area is finished. This takes time, of course, and for this reason the systems are appropriate only for woodlots with good concentrations of wood ready for harvest. Since all three systems—the chutes in particular—have limited lateral range, you will have to do some hauling if you wish to avoid constantly breaking down and resetting. Note also that two of the systems (suspended cable and chutes) require a fairly sharp downhill skid (25-70% grade is best) to work effectively, and that the third, the ground cable system, is most effective when the wood is to be pulled uphill. In short, though these yarding methods can work in terrain unsuitable for vehicles or animals, they are at the mercy of terrain to work at all.

Another feature of exotic skidding systems is that they require considerable patience and tinkering ability. Since you're working in concert with the landscape rather than in spite of it, no two settings will be the same with any stationary system. This means a fresh set of problems to solve every time you move, and it means that you'll quickly learn to keep your pockets full of wire, baling twine and duct tape and your mind full of enthusiasm for dealing with unforeseen challenges.

If you have access to an appropriate woodlot and your goal is a responsible wood harvest with a relatively small outlay of cash and a relatively large outlay of time and energy, then read on. And don't forget: If you can get one of these outfits working properly, it will probably produce as much fun as it does wood.

The Ground Cable System

The name is mine, but the system is the creation of Tom Bahre, a former Addison County, Vermont, forester who devised it in the '70s to provide noncommercial cutters with an inexpensive way to harvest off-road wood.

"It looks pretty silly when you compare it to, say, tractors," says Tom. "But when you've got a bunch of wood you can't get at any other way, it doesn't look so silly anymore."

Described very simply, the ground cable system involves blocking

FIGURE 50. *Tom Bahre's capstan assembly*

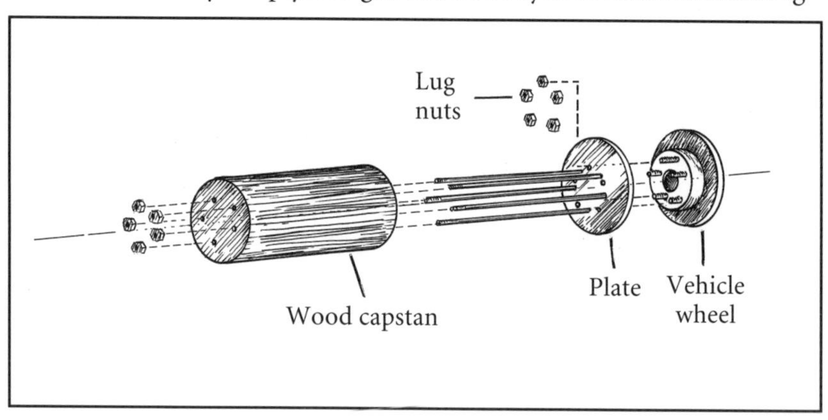

FIGURE 51. ***Tensioning the capstan***

up the drive wheel of a vehicle, replacing the tire and rim with a modified hub and attached capstan, stringing rope from the capstan to an anchor tree in the woods and attaching an old car hood to the rope with a swivel hook. Once you've loaded the car hood with wood and chained the load down, simply put the vehicle in low gear and haul in the goodies. To return the hood to the cutting area, put the vehicle in reverse.

Tom had to do a fair amount of modifying to make the system work. What he came up with was a 3/16-inch steel adaptor plate that bolts onto the vehicle's hub (using lug nuts and bolts) and has long bolts welded to it to secure the capstan, a 1-foot-long piece of sugar maple that he had turned and drilled at a woodworking shop (FIGURE 50). Before attaching the capstan, Tom would position his truck with the front bumper against a tree to keep it from moving, and with another tree behind it where he could set up a come-along and pulley to tension the capstan (FIGURE 51). Without the tensioning, the rope would only spin around the capstan instead of winding tight and pulling in the load.

As mentioned earlier, this system is best suited for skidding uphill, because gravity helps to keep it under control. At the crest of any hill, Tom found it necessary to run the rope through two pulleys rigged off the ground in a tree, one pulley for the return side of the rope—approximately five feet off the ground—and the other pulley for the load-bearing side of the rope, set a foot or two below the first one (FIGURE 52). Tom made these pulleys by welding an old ball joint to a steel plate and attaching a tire rim. He found that the rope

FIGURE 52.
Pulleys at the crest of the hill

would often hop out of the bottom pulley as the load passed, so he welded steel rebar fingers to the edge of the rim to keep the rope in place. He also staked a bumper log under the pulley tree to push the load aside so that it wouldn't hang up as it passed.

FIGURE 53 shows the whole system. The "cable" in Tom's ground cable system is a misnomer, as he used 700 feet of 1-inch manila rope, giving him a working distance of 350 feet. He used rope because he happened to have it lying around and because the manila resisted the considerable heat generated by friction on the capstan.

The strengths of the ground cable system are that it is relatively inexpensive and quite fast: It will move several hundred pounds of wood faster than you can comfortably walk, works well over wet ground and does very little damage to the ground or residual trees. Tom also notes that by rigging a snatch block and tying a smaller rope into the system, he was able to pull wood to the skidding corridor from the side. The drawbacks are the setup time, which can run to several hours, and the system's need for specific terrain.

The Wood Chute System

"If you can find another way to get your wood out, use it," says Becki Bates of the firewood chute system she designed and built to carry out a timber stand improvement project in central Vermont. In her case, the woodlot that needed thinning was situated on a very steep (35% grade and worse) slope with many outcroppings of ledge; terrain so inhospitable that walking to the trees to fell them was difficult (I know, because I did a lot of it), and wheeled or tracked vehicles and draft animals would have had virtually no access.

Becki considered a manufactured wood chute system, but dismissed it as far too expensive (thousands of dollars for the amount she needed) and decided to build her own out of 16-inch pvc pipe. To do this, she sawed each 8-foot section of pipe in half lengthwise with a circular saw and drilled two holes in each end. These sections, along with copious supplies of wire, baling twine and rope, were trucked to the foot of the hill and carried by hand to the top. The sections were joined to make a continuous chute by attaching them with wire through the predrilled holes. The top section was secured with a rope to a tree, and the chute was anchored to trees along the side as the terrain demanded—particularly where it was necessary to angle the path. Note that angles are to be avoided, though; even a 10-degree divergence will often cause the wood to jump out of the track. Also, any place where the chute, because of a change in the slope, was not firmly on the ground, it was supported with brush, snow or wood to keep it from shifting and breaking.

To operate the system, simply chuck your wood into it and let gravity do the rest, which it will with breathtaking speed. So much speed, in fact, that no person, animal or breakable equipment should be near the bottom of the chute or near any angles or grade changes where wood is likely to jump out. Obviously, the wood must be at least slightly smaller in diameter than the chute, so larger chunks have to be split in the woods. Becki also found that two feet was about as long as a stick of wood could be and still make the journey to the bottom, so the bucking must also be done in the woods.

FIGURE 53. *Complete ground cable system*

FIGURE 54. *Wood chute construction*

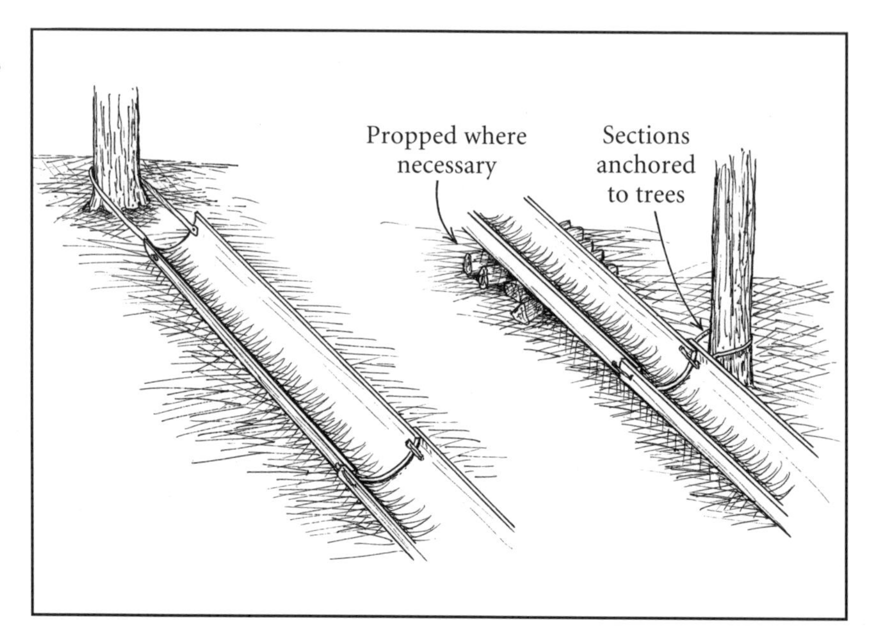

Other problems included log jams when one or two pieces would stop for some reason (size, branch stubs, etc.) and back up everything that came after, breakage of wire and baling twine connectors and guy wires, and shifting of sections of the chute—all of which resulted in down time and frequent frozen-fingered tinkering. Wet snow in the chute, mercifully infrequent, proved to be so effective a brake that the whole operation would have to stop until the pipe was cleared. These problems would certainly be minimized if your chute was short enough so that you could see its entire length from the top and could thus avoid feeding wood into it while it was breaking up below.

I've dwelt on the problems, but I'll close on a positive note: We moved many cords of wood—none of which was accessible using other skidding methods at our disposal—500-600 feet through the chute with little or no environmental impact and had a lot of fun doing so. FIGURE 54 shows the chute.

Suspended Cable, or High-Lead System

For years, some of the biggest timber in the world was yarded by the high-lead method, which made it possible to harvest the otherwise inaccessible timber in the rugged coastal ranges of the western United States and Canada. This type of logging, featuring stationary winch engines powering suspended cable through pulleys rigged 100 feet or more up the trunk of a distant spar tree, was considered most practical for clearcutting and, because of the time and difficulty involved in moving the engine to a new location and rigging a new spar tree, economically feasible only for large volumes of high value timber.

The high-lead system's intrinsic advantages over other skidding methods include the wood's being completely off the ground while it is being moved, so wetlands and rough terrain don't have to be avoided. Also, soil and residual timber damage are reduced or eliminated, and costly and time-consuming road construction is minimized. These advantages make the system very attractive to anyone interested in low-impact woodlot management, but the cost of the equipment and the time and skill required to set it up have precluded its use for a small scale firewood thinning operation.

Enter E. Gerry Hawkes, a forester and firewood producer from Woodstock, Vermont, who, recognizing the need for economical, low-impact skidding, developed several gravity-powered, suspended cable systems. The simplest consists of a single 3/16-inch high-tensile-strength steel wire drawn taut between a tree at the top of a slope and one at the bottom, as much as 1,600 feet apart. The harvested wood is hooked to the cable by twine or wire looped on a sliding hardwood block and is then released, whereupon it slides to the bottom of the hill. Gerry usually guys an old tire to the line at the bottom to serve as a stop for the load; the wood hits the tire, the impact breaks the wire or twine and allows the wood to drop. Concave slopes with a 25-70% gradient are best suited to this system (FIGURE 55).

The main disadvantage of this simple high-lead system—that wood can be hooked on only at the top where the wire is close to the ground—is eliminated in Mr. Hawkes's more sophisticated version which employs a vehicle and pulleys. Instead of securing the wire to the tree at the foot of the hill, you pass it through a pulley which is secured to the tree, and tie it to a truck. By backing up, the truck driver slackens the cable, allowing wood to be hooked to it at any point in the skidway. By driving forward, the driver tensions the cable, causing the load to slide downslope (FIGURE 56).

Either suspended cable yarding system offers two distinct advantages over the chute and ground cable systems: 1.) Because the wood is off the ground in transit, deep snow and rough ground have little if any effect on it; and 2.) It is able to cover a longer distance.

The disadvantages are the same as for the other two fixed-in-place systems (setup time, terrain requirements), but with the addition of one more: Somebody will have to climb the trees to rig the wire and pulleys—not too high (20-30 feet), it's true, but high enough to discourage those sensible folks who like to feel the earth beneath their feet at all times.

FIGURE 55. *Simple high-lead system*

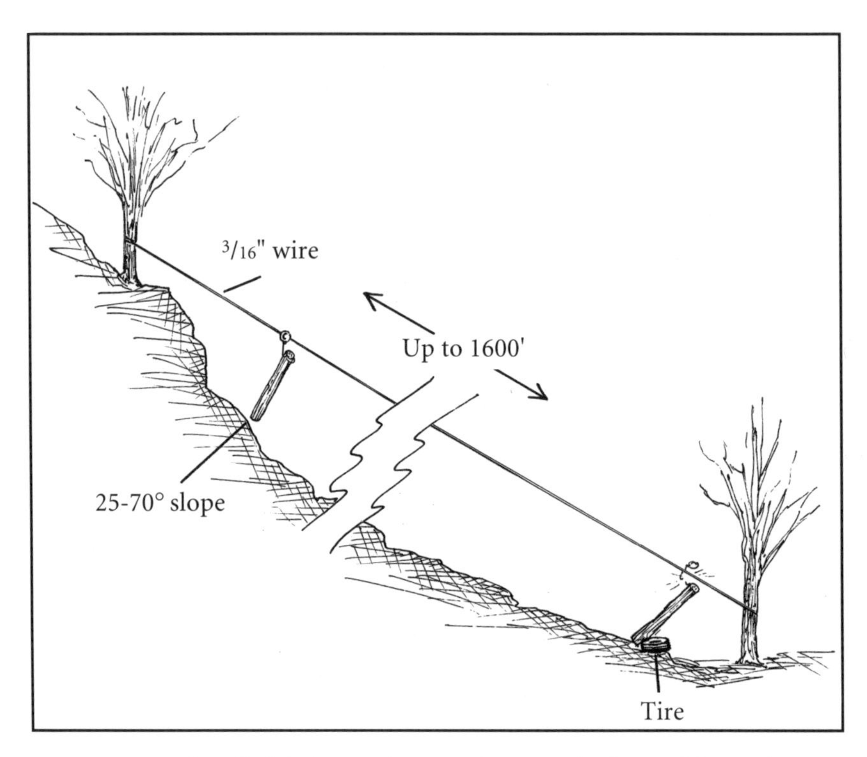

For more detailed information on these and other cable yarding methods, I refer you to *Cable Harvesting Systems for Small Timber,* by E. Gerry Hawkes, published by the Vermont Department of Forests, Parks and Recreation, Waterbury, Vermont 05676.

BUCKING

When I think about bucking—cutting wood to the desired length—I'm reminded of the sword swallower in a small circus that came through town when I was a kid. He opened his act by saying, in a hoarse voice, "It's easy, when you know how to do it," whereupon he swallowed an impossibly large broadsword blade. Well, bucking is easy when you know how to do it, but like sword swallowing, it can be frustrating and dangerous if you don't, or if you let your attention lapse regardless of your skill and experience.

The frustrations to which bucking may expose you consist mainly of getting your saw pinched in a cut or dulling your chain when you cut the dirt accidentally. The dangers include saw kick-back and other kinds of cuts as well as heavy pieces of wood rolling and/or dropping in your vicinity. We'll begin by listing and discussing some general safety rules for bucking and follow with procedures for making different kinds of cuts. Following these rules, practicing and staying alert

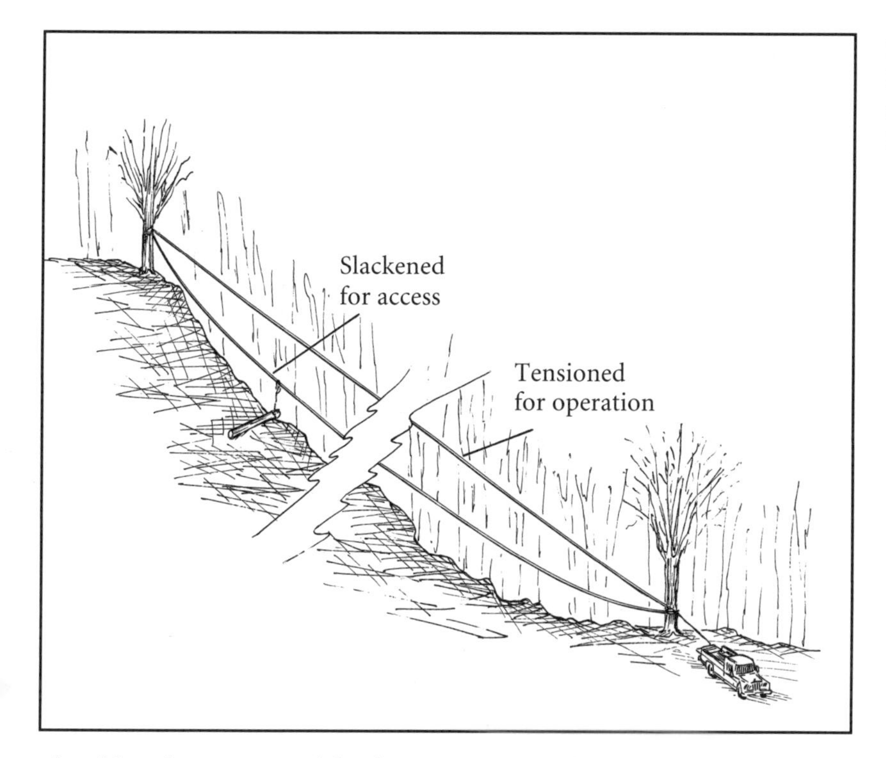

FIGURE 56. *High-lead system using truck for tensioning*

should make your wood-bucking career as safe and uneventful as a day in the life of a knowledgeable sword swallower.

I'm assuming, as I do throughout this section, that the cutting tool being used is a chain saw. If your choice of tools is a bow saw or two-person crosscut, you'll find that, because of the size of your saw and its inability to make a plunge cut, most of the techniques I describe for dealing with compression and tension will be unavailable to you. Instead, use a bucking wedge to counteract the pressure, and you should get along just fine.

Safety Rules

1.) *Use a firmly balanced stance at all times,* your feet approximately shoulder-width apart and nearly perpendicular to the log you're cutting, rather than parallel (FIGURE 57). In this position, you can cut with your saw held to one side so that you aren't directly behind it in the event of a kick-back, but you'll still have it close enough to your center of gravity to control it comfortably.

2.) *Two hands on the saw always.* This seems obvious, but many an experienced sawyer will take his or her front hand off the handlebar to scratch, puff a cigarette, or just show off. Such nonchalance can be fatal.

FIGURE 57.
Balanced stance for bucking

3.) *Keep your work area clear.* The most solid stance in the world won't save you if you trip over the chunk of wood you just cut, and falling with a running chain saw in your hands is more excitement than most of us need.

4.) *Spare the throttle, spoil the sawyer.* It's best to start your cuts with the saw at full, or nearly full, throttle; kick-backs can occur when the chain is not turning rapidly enough and the teeth are stopped suddenly in making contact with the wood instead of cutting. Another unpleasant sort of accident sometimes associated with slow chain speed occurs when the chain pulls the wood forcefully against the body of the saw, sending the stick flying, often at the sawyer. This doesn't mean that you should walk around with your saw revved to the max; it means that you should accelerate smoothly into each cut, depressing the throttle just before the chain makes contact with the wood.

5.) *Keep your bar tip in the clear.* Kick-back is always a possibility when you are using a chain saw, but particularly so when you are bucking wood in a pile, such as a log truck load or long wood bunched up on a landing. This is true because the pile provides a wealth of potential obstructions with which your chain can come in contact as it goes around the upper tip of the bar. When this happens, the chain may suddenly stop turning, causing the force generated by the still-revving engine to throw the saw backward, out of the cut. To avoid this unpleasantness, always pay attention to where the tip of your bar is. You may find it prudent to use your peavey to roll some logs free of the pile before bucking, and this is not a bad idea from another standpoint: By taking the pile apart in a controlled way, you can avoid having it fall apart unexpectedly when you make the wrong cut and are standing in the wrong place.

6.) *Remember gravity.* When bucking on a hillside, stand on the uphill side of the log, in case it doesn't stay put when you cut it.

Bucking Small-Diameter Wood

Just as you wouldn't learn to drive behind the wheel of a Greyhound bus, you shouldn't begin your education in bucking by tackling a 4-foot-diameter tree. Starting small gives you the opportunity to get a feel for your chain saw and for how to cut wood on the ground. What is small? For our purposes we'll say 8-10 inches in diameter and smaller. Wood this size can usually be bucked with one cut rather than a sequence of cuts, and can be easily rolled or propped on another piece of wood or a saw horse to minimize the number of times you cut rocks and topsoil.

The procedure, at its simplest, goes as follows: Assuming that the piece of wood you wish to cut is propped up—supported at one end only—take the proper cutting position and cut straight through from the top down. Keep your feet out of the way.

If the wood you wish to cut is supported at both ends, either between two logs or between a log and the ground, follow the same procedure, but cut from the bottom up, since the compression in this situation would cause a top-down cut to pinch the saw.

If the wood in question is lying flat on the ground, supported along its entire length, you can either prop it up or cut it part of the way through, roll it 180 degrees, and finish the cut. This is a technique you'll need anyway, because there comes a point when the log you're bucking is so short that the saw will roll it instead of cutting it. When you reach this point, you'll find that you can cut the log much more easily when it is flat on the ground.

Bucking Large-Diameter Wood

As you graduate from bucking small wood to bucking larger logs, you'll need to add some new skills to your repertoire. The greater weight of large logs and the greater time it takes to cut through them increase the effects of pressure and gravity. For example, when you are cutting a 6-inch-diameter stick that is supported at one end, you can simply make one cut from the top down. Increase the diameter of that stick to 16 inches, however, and you'll often find that the wood begins to split and bind your saw before you can complete the cut. To avoid this problem, first cut from the bottom up (called "underbucking") a quarter to a third of the way through the log (FIGURE 58). Now cut from the top down to meet the underbuck. This two-cut method drastically reduces the likelihood of splitting and binding. If the log in question is supported at both ends, simply reverse the sequence of cuts; cut a quarter to a third of the way through from the top down, then finish with an underbuck.

If the log is too close to the ground to underbuck—if, for example, it's supported at both ends by the ground instead of by other logs—you can either try rolling it with a peavey or using a wedge to hold the cut open as you buck from the top down. You've probably already discovered that logs on the ground are subject to pressures that you can't always identify accurately. You'll make one cut without any difficulty and bind your saw as tightly as the sword in the stone on the very next cut, for no apparent reason. This is why it pays to be a little tentative when you aren't certain of the pressure, and pays also to keep a wedge or two handy.

The step from large logs to logs greater in diameter than your saw's bar is long is a large step indeed, because it will lead you to sequential cuts that involve a plunge cut, and a plunge cut poorly executed is a gold-plated invitation to the kick-back gremlins.

If a very large log is clearly supported—not resting flat on the ground—you can probably get away with using the two-cut method, first on one side, then the other, as in FIGURE 58. It gets tiresome walking around the log, however, and at some point you'll find that

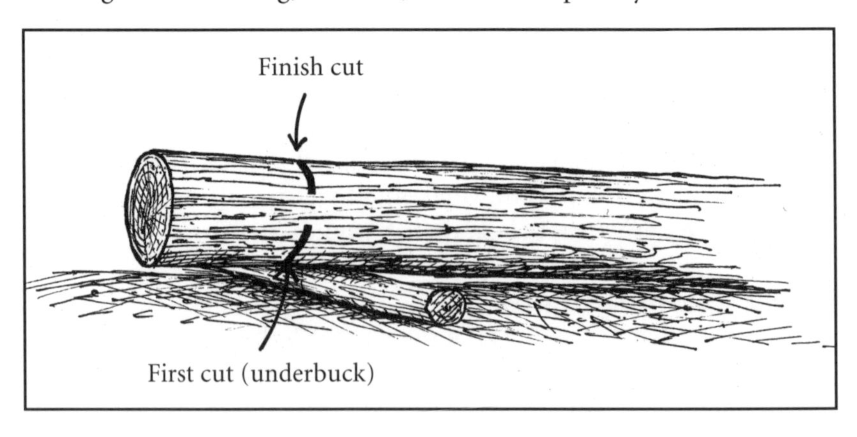

FIGURE 58. ***Two-cut method for bucking large-diameter logs***

the log is too close to the ground to underbuck and too heavy to roll. So here's the sequence if the log is supported at one end (FIGURE 59):

Cut Number 1 is simple enough; just remember to bend your arms and knees enough so that, as the cut progresses, you're still cutting with the bottom of the bar rather than the tip.

Cut Number 2 is the plunge cut, assuming that there isn't room for a regular underbuck. The trick here is to hold the saw's powerhead up so that the top of the bar is cutting, not just the tip (FIGURE 60), and to push the tip slightly up and in. Remember that slow chain speeds invite kick-back, so give your saw plenty of gas. If you do this correctly, much of the thickness of the log will be between you and your chain should a kick-back occur. What you'll feel will be a powerful jolt or series of jolts which will tell you that you're holding the saw at the wrong angle, not giving it enough throttle, or both. With experience you'll be able to make the cut smoothly and uneventfully, but you need to be very comfortable with your saw before you begin to learn the plunge cut.

The third cut in this sequence is made by reaching over the log (FIGURE 61) or by walking around it if it's too big for you to reach over and still retain firm control of your saw. (Obviously, with either this or the two-cut method, if you have to walk around and cut the other side of the log, do it just once: Do all of the cutting on one side, then go around.) Make certain that cut Number 3 meets cut 2.

Cut 4 is self-explanatory and simple, as FIGURE 59 shows. The only thing to look out for is the possibility that you misread the pressure, in which case the cut will bind. Be tentative until you see the cut begin to open.

When cutting a large log supported at both ends, just reverse cuts 2 and 4 in the sequence (FIGURE 62), although I usually reverse 1 and 3 as well (FIGURE 63), because it's a smoother progression—you are, in effect, cutting in a circle around the log.

A log too large to roll that is lying flat on the ground will give you another opportunity to use the plunge cut to your advantage. The sequence of cuts will be the same as described. The only difference is in the execution of cut Number 2, the plunge cut, and the only difference in that cut is how slowly and cautiously you finish it, since going too far will bring your chain into contact with the ground. After making cut 1, start your plunge cut and push the tip of your bar through to the other side, if it will reach. (It may, because you aren't making this cut at the widest part of the log. If it doesn't you'll have to work from one side, then the other.) Bring the bar up as you cut until it's parallel to the ground and approximately one-third of

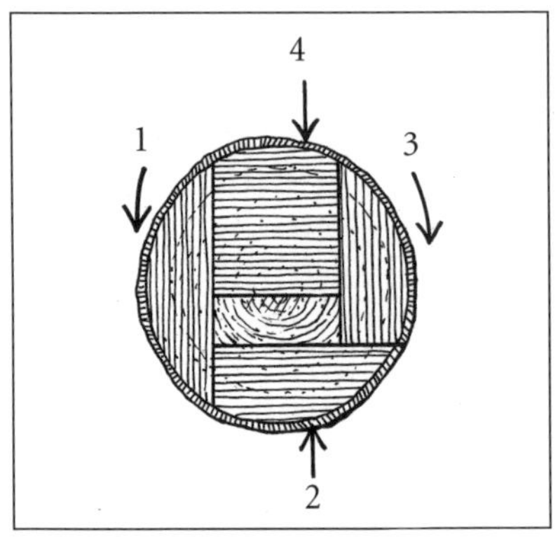

FIGURE 59. ***Sequence of cuts for bucking very large log supported at one end***

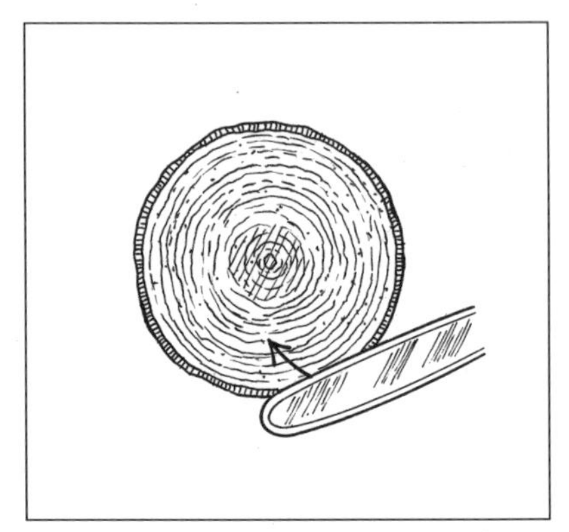

FIGURE 60. ***Begin plunge cut with top of bar.***

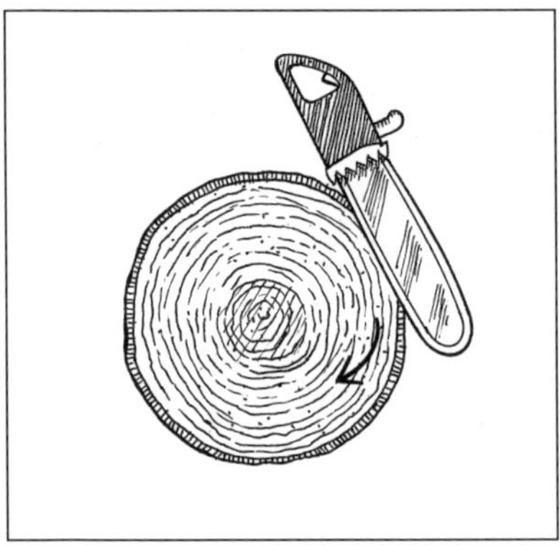

FIGURE 61. ***Cut reaching over log.***

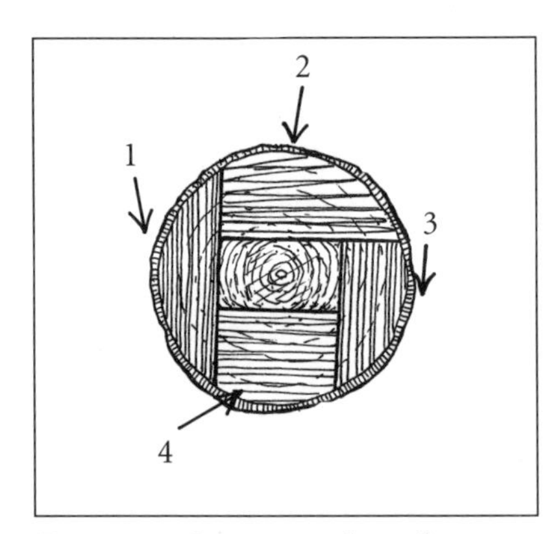

FIGURE 62. ***Sequence of cuts for bucking very large log supported at both ends***

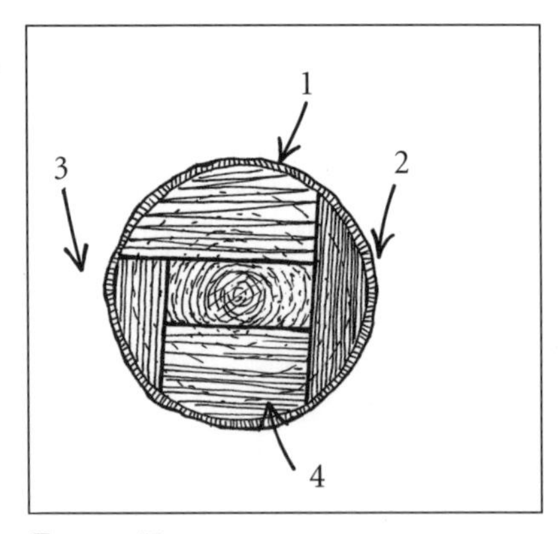

FIGURE 63.

the way through the log. Now slowly work the bar back down toward the ground, stopping just before you cut through the bark, but after you've cut all of the wood. This is a good trick, and you won't manage it every time. Now go on to cuts 3 and 4 unless you goofed and have to file your chain first.

Measuring and Marking

My stove will accommodate a stick of wood 18 inches long and not an inch longer. This leads to a lot of measuring when I'm bucking a good-sized tree, so how do I do it? I have a 20-inch bar on the chain saw that I generally use, and the bumper spikes protrude almost 2 inches from the saw's powerhead, so when I'm bucking, I simply turn the saw sideways to measure and then cut a piece a hair shorter than the distance from the tip of the bar to the spikes. Except when I'm cutting small-diameter limbs: Those cuts I just eyeball. After all these years, I'm quite accurate.

So how does my method work? My percentage of accurate cuts is pretty impressive—maybe 98%. A .980 batting average would probably guarantee me a $25 million annual major league contract, even if I couldn't field. But we aren't talking about doubles in the gap, we're talking about wood in the stove. Many of my mistakes are 17 or even 16 inches long. Not a big problem; this just means that there's more air space than necessary in the firebox. Sometimes, though, my mistakes are 19 or 20 inches long. Some of these I spot as I'm taking them off the woodpile, and they join the growing mound of rejects which I will or will not shorten to 18 inches someday. A few of the 19-inchers make it into the firebox. I can never believe I won't be able to make them fit, so by the time I'm done poking and swearing and admit defeat, they're burning. I have to run outside, grab the metal ash bucket, pluck the burning, smoking stick from the stove, and carry it through the house with familial admonitions about sparks on the rug ringing in my ears and fear in my heart at the prospect of tripping over a large dog sleeping in the hall. Once outside, I fling the offending stick into the snow. It will reappear, along with its nonconforming kin, with the arrival of the spring thaw.

In short, my method works fine, if you need a little comic relief with your firewood, which I apparently do. Not so fine, otherwise. This is why I encourage all buckers to measure and mark before they cut. You can measure with a tape, of course, or you can make a measuring stick (a "wand," as some loggers call it). Just be sure that the wand is the right length. You'd hate to go through my little ritual if you'd measured, wouldn't you?

Marking is most easily done with lumber crayons, which are made for the purpose in a variety of colors and are available through most chain saw dealers. These have the added benefits of turning your fingers and pockets whatever color they happen to be and of getting lost frequently. You can also mark your cuts with a blow from a sharp axe or hatchet, but it's hard to be consistently accurate this way, and you know what might happen if you aren't accurate.

Other Helpful Tools

We've already mentioned peaveys several times; they're handy to have around any time you are working with logs that are too heavy to lift.

Plastic wedges, and something to drive them with, are also good to have when you're bucking; they can save you from the special frustration that only a stuck chain saw can cause.

Our old friend the pulp hook, while not a necessity for bucking, comes in handy for clearing cut pieces out of the way and for pulling logs off a pile. Even better for this last job is a truly evil-looking tool called a pickaroon (FIGURE 64), which allows you, courtesy of its long handle, to be farther out of harm's way when the logs come bouncing down.

That ought to be enough to get you started at bucking. Remember, be careful with the plunge cuts, measure and mark before you cut and, always, keep those lumber crayons dry.

SPLITTING WOOD

If I were of a philosophical bent, I would say that the subject of splitting wood is a microcosm of the larger subject of woodburning, including as it does an element of old ways versus new ways, along with a healthy dose of folklore. The old way is doing the job with hand tools, and the new way is doing the job with a power splitter. The folklore is clearly attached to the old way; hydraulic splitters, whatever virtues they possess, don't inspire much myth-making, study of natural phenomena or refinement of technique.

I may as well tell you now, because you'll figure it out soon enough, that I am firmly in the split-it-by-hand camp. The reason is probably because I am a spiritual descendant of the Luddites, innately hostile to new ways of doing things. I can also give you four more reasons for my sympathies, reasons that sound a bit more open-minded than the first: 1.) Hand tools are cheaper and quieter than power tools; 2.) I like swinging a maul: It is satisfying and is good exercise; 3.) As a part of my preferred method of working up firewood—buck a little, split a little, stack a little, buck a little, split a little, stack a lit-

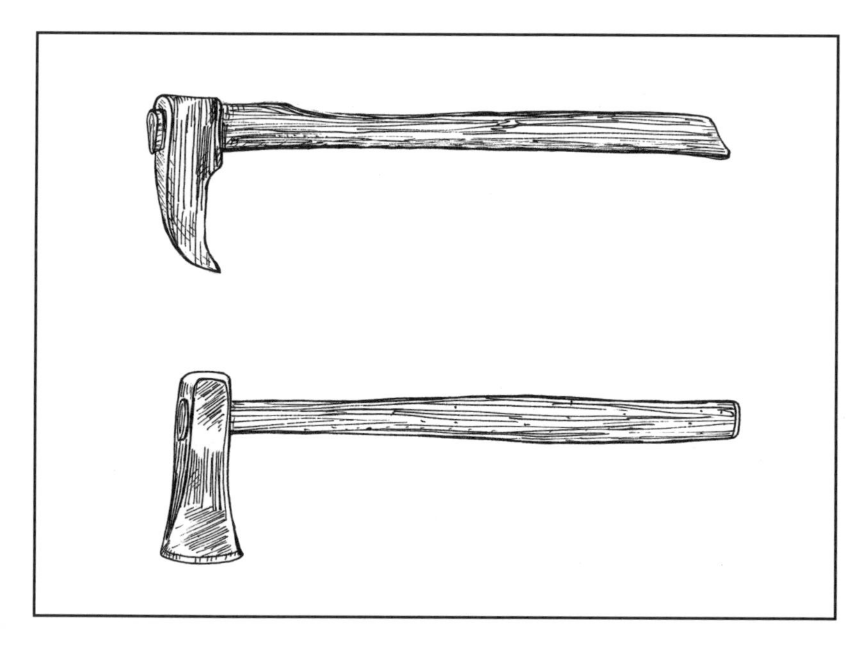

FIGURE 64. *Pickaroon and splitting maul*

tle—splitting by hand is nearly as fast, if not as fast, as splitting with a machine would be. I don't have to start my maul, and I can carry it to the wood, rather than vice-versa. I have, on occasion, split by hand alongside a hydraulic splitter, and I've definitely kept up with the machine. Friends of mine who use splitters point out that I'm no John Henry, which is true enough, and claim I would never keep up with the machine for a full day. That may be true, too, but it may not: The machine might not tire (though it might break a hydraulic hose or refuse to start), but the person operating it would: Lifting and carrying wood to a splitter is hard work. I'll concede this much: If the wood to be split is knotty or is, God forbid, American elm, then a splitter will beat a maul every time. If the wood is straight-grained, forest-grown and not enormous, then I'll reserve judgment, and reiterate that I don't split wood all day, anyway.

Reason Number 4 for splitting by hand relates to folklore: Part of what I like about cutting and burning wood is that it puts me in touch, in a small way, with the independent people who came before me, gives me a chance to learn things that they knew and, maybe, view life and the passage of time in a manner similar to theirs. When I split wood, it is a means to an end, but it's also a bit of an end in itself. Splitting wood with a splitter seems to me to be just a means to an end—hurry up, hurry up!—and is uncomfortably like a lot of other conveniences that mainly serve the purpose of buying us a little more time to watch television.

You, of course, will have to decide which way of splitting wood suits you best, and in fairness to power splitters, I'll point out that I do, inevitably, find myself splitting big, knotty chunks of wood by hand when the machine would work better, and also that I'm the only person in my household who can realistically be expected to split a winter's supply of wood, while my wife and children could certainly do the job with a splitter.

Hand Tools

If you choose to do your splitting with hand tools, you'll have another choice to make, because there are four distinct ways of doing the job, plus gimmick tools which appear and usually disappear from time to time. A very traditional and appropriately named tool for splitting is the splitting maul (FIGURE 64). Most commonly available with a 6- or 8-pound head (heft both before choosing), these versatile tools have a fairly sharp edge for splitting and a blunt side for driving splitting wedges when the sharp side doesn't do the job. Splitting mauls aren't terribly heavy, but it does take some strength to swing one effectively, particularly when a large, tough piece of wood requires a few hard swings to be split.

The maul's wooden handle is its weak point: Like the clutch in a driver's-education-class car, wooden maul handles get sacrificed to the learning process. Even when you get proficient enough to eliminate those disastrous overswings, you'll still break a maul handle from time to time: Weather and the repeated shocks of the work it does will weaken the wood. Get someone to teach you how to hang a new handle—unless, like a certain nameless government agency, you conclude that it is really cheaper to throw the whole maul away and buy a new one. I feel certain that anyone choosing the latter course of action will pay for it in the end.

A sledgehammer and three or four large steel splitting wedges will eventually split almost any piece of wood worth splitting. The hammer and wedges are generally not as fast as a maul, but if you routinely deal with knotty, twisted-grained wood, or you are splitting for a furnace that takes sticks 3 or 4 feet long, they are probably the best tools to use. Two notes of caution, however, and since maul users will have to drive wedges from time to time, they too should take note:

1.) When you hit a steel wedge with a steel hammer, chips of metal occasionally go flying. Eye protection is strongly recommended. 2.) Never drive your last wedge so far into a piece of wood that you can't knock it back out; you can use one wedge to retrieve another, but not if they're both buried to the hilt. Your only recourse is to go to the store and buy a third wedge, and if you find yourself doing this repeatedly, you'd do better to buy a hydraulic splitter.

A third splitting tool, the so-called Monster Maul, looked to be a gimmick tool when it first appeared in the '70s, but it has clearly stood the test of time and is, in fact, my tool of choice for splitting wood. A Monster Maul differs from a regular maul in several key ways. The angle of its cutting edge is more obtuse, so it seldom gets stuck in the wood. The handle is steel, so it seldom breaks. (Not never, but seldom, and when it does, you either get it welded back on or use the government solution.)

A Monster Maul is heavier than a conventional maul. One model had a 16-pound head and a total weight of 20 pounds, and the more common size has a 12-pound head and a total weight of 15 or 16 pounds. This weight makes a Monster Maul harder to swing than a conventional maul, but because it's so effective, you don't have to swing it as often. There are lighter—6- and 8-pound, I believe—versions, and they probably work well, but if you can handle it, the heavy maul generates the more forceful impact. (Even though a Monster Maul has a blunt side in addition to a cutting side, anyone who has ever had his or her hands and arms numbed by one knows that this tool, because of its hollow handle, isn't much good for driving wedges.)

Last but not least, the Monster Maul is much uglier than a conventional maul. The makers have tried to pretty them up with rubber hand grips and good paint jobs—the originals were simply dipped in orange machine paint—but they're still ugly. I know that form follows function, but this gives me pause.

Some people prefer to split their wood with an axe. I suppose that if you mostly split very short lengths of straight-grained wood—say, for an old cook stove—a nice 3½-pound axe would do a fast and easy job. Otherwise, because axes are light and have thin, gradually tapered cutting edges, what often happens when you try to split a chunk of wood with one is that it gets stuck—really stuck—and you use a great deal of energy and a large number of good words trying to retrieve it. If you are going to use an axe for splitting, don't use a double bit (an axe with a cutting edge on each side). Sometimes, when you hit a piece of wood just wrong, the tool you hit it with bounces right back up at you. This is bad enough when the side nearest you is blunt steel; when it's a sharp edge, it can be a really unpleasant experience.

A general note about splitting tools that are swung: Accuracy is, for all practical purposes, the same thing as both safety and effectiveness; if you always swing your maul accurately, you will split more wood with less effort and will not have to contend with hard things bouncing off your shins and feet. It takes practice to be accurate, and it takes proper technique: When splitting wood with a maul, axe or hammer and wedge, your swing should come straight over your head, not angled over your shoulder. I find that it helps to slide my top hand toward the tool head on the upswing and slide it back to my other hand on the downswing; this distributes the weight and adds guidance and leverage.

Gimmick Tools

"Gimmick" may not be a fair word to apply to the odd assortment of unconventional splitting tools that have come and, in many cases, gone through the years. With the exception of the Monster Maul—which is, after all, still a splitting maul—none has caught on. Whether this is because they often don't work very well and tend to be expensive, or because the tried-and-true equipment works well enough and the people who use it are—as I am—resistant to change is a question I can't answer. In any case, I'll mention a few of the innovative splitting tools I've seen in case any of you folks are adventuresome. Somebody needs to check new ideas out, or we'll all miss that better mouse trap.

"Chopper" axes are regular single-bit axes with spring-loaded levers on the sides of the blade. The idea is that, as the axe blade sinks into a stick of wood, the levers pop out sideways, forcing the wood to split. I've taken a few swings with a chopper and have known people who swear by them. My reaction is that what they do more effectively than a maul is send the split pieces flying; not an advantage unless you like the idea of doing a little walking to collect the fruits of your labor.

Splitting bar. I'm not certain what to call this, but it looks like a large pry-bar with a steel wedge on the end. A version I saw advertised recently has a weight that slides up and down the handle, and I seem to recall an earlier model with a heavy, sleeved outer handle that traveled up and down the inner handle. The idea is to position the bar on the end of the piece of wood and slide the weight device forcefully into the wedge. The manufacturers point out that this tool is easy on the back because it eliminates swinging and is safer because it eliminates being hit by poorly aimed swings and flying metal chips. They also say it works. I don't know, personally, but if you have back trouble, you might want to check one out. Still, I'd ask a few questions of someone who has actually used one before I plunked down the $60 or so that these rigs fetch.

Funny wedges. I've seen two different ideas for changing the time-honored shape of the splitting wedge for the purpose of making it more effective, and the concepts are appealing for their simplicity: No bells, whistles or hinges; just a hunk of steel molded in an innovative shape. Sometimes great ideas are so simple that they elude everyone for years, right? Well, I actually tried one of these wedges some time ago when, I suppose, I was less jaded than I am now. It was called a Wood Grenade, and I don't know whether or not it's still available. I haven't seen one—including mine—in years. The grenade was conical and tapered to a sharp point which was threaded—almost barbed, really—to keep it from popping out of the wood. The idea was that, because of its shape and thickness, when you drove it into the middle of a piece of wood, the grenade would not only split it easily, but would often split it into three or four pieces. Eagerly I carried the grenade to the woodpile, along with my sledgehammer, and eagerly I put it to the test: The first time I hit it, it went perhaps half an inch into the stick of elm. (I was being a tough judge.) No split. The second time I hit it, the grenade popped out of the wood, did a neat somersault, and landed at my feet. A good trick had it been a dog. Maybe elm was an unreasonable test, I thought, so I tried it again on an ordinary stick of sugar maple with the same results. It was getting dark, so I gave up, but every once in a while, suspecting perhaps that I'd imagined the earlier failures or that they were somehow my fault, I'd try the grenade again. A few times it actually split a piece of wood, and it even, on one occasion, split a perfect, straight-grained piece of white ash into three pieces. (A stern look might have done as much.) But one day, when I was trying it yet again, it went popping into the air with more than its usual vigor and disappeared into the tall grass. If somebody owns my property someday who is more enthusiastic about expanding lawns than I am, he and his mower will probably find it.

The other funny wedge design looks like a conventional splitting wedge that has been twisted a half turn or so, the idea being that the corkscrew shape will generate force as the wedge penetrates the wood. It looks reasonable, but so did the Wood Grenade.

Splitting Tips

Now it's time for folklore. Over the years, people who split wood by hand collect tidbits of information that help them get the job done. It doesn't matter that their neighbor's tidbits are directly contradictory to theirs, so long as everybody is happy and the wood gets split. I'm going to give you tips I use: Some you will probably decide are wrong, and some you will, I trust, find useful; they all work for me. I'm not going to discuss which phase of the moon is best for splitting or what omens to look for before shouldering your maul,

FIGURE 65. *Hydraulic splitter*

but some of my hints, given the variability of wood, people and tools, would be just as hard to document.

1.) Wood—except for beech—splits easiest when it's freshly cut.

2.) Wood splits best when you strike the most recently cut end. (A few days' difference in when the cuts were made won't make much difference, though.)

3.) If a piece of wood has branch stubs, split the end that points in the direction opposite that of the branches.

4.) Wood splits easily when it is frozen. The colder the better.

5.) Use a chopping block. It will save the edge of your maul and doesn't give and diffuse the force of the blow as the ground will.

6.) If you are plagued by pieces of wood tipping over when you try to split them, get an old tire (small ones work best) and stand your wood in it.

7.) If you use a maul or axe, don't keep smacking a piece of wood that obviously doesn't want to split unless doing so gives you perverse pleasure. After a predecided number of futile blows (the number being dependent upon how much time you wish to waste), use a wedge or rip the wood with a chain saw. There's little shame in doing this occasionally.

8.) Try to avoid splitting directly through knots.

9.) If you use a hammer and wedges, keep an axe handy, preferably a nonprecious one. It takes a long time to sever all the strands in a stringy piece of wood with wedges, and an axe can usually accomplish this with a few short strokes. Of course, you'll hit the ground sometimes; that's why you're using an old, beat-up axe.

10.) If a piece of wood shows no sign of splitting, try the other end—it often works as an alternative to wedges and rip cuts.

11.) It is usually easiest to split a piece of wood at the end farthest from the knots.

12.) I've saved the safety tips for last: We mentioned eye protection; steel-toed boots and even shin guards can save you some pain. Also, keep helpers and onlookers a safe distance away, because wood, metal chips and even tools go flying. Speaking of flying, check regularly to be sure that tool heads are securely joined to handles. If a tool head is loose, don't use it until you've fixed it, either by adding handle wedges or soaking it overnight, or both.

Power Splitters

By far the most common power splitter is a gasoline-engine hydraulic model. These are simple machines, consisting of a heavy steel frame, a small 4-cycle engine, a hydraulic pump and a steel wedge and plate, one of which is mounted on the shaft moved by hydraulic pressure (FIGURE 65). When a piece of wood is set on the tray and the handle that activates the pump is pulled, the shaft pushes the wedge into the wood which is held in place by the stationary plate, or, depending upon which way the machine is set up, the shaft with the attached plate pushes the wood against the stationary wedge.

Variations on this theme include splitters that use electric motors rather than gasoline engines, and one model that used a fly wheel in place of hydraulics to increase the speed.

If you suspect that a power splitter would suit you better than a maul would, be aware that you'll probably have to spend $1,000 or more to get a new one capable of serious year-in, year-out work. Here are a few things to look for when you go shopping for a splitter:

1.) Look for a machine with a two-stage hydraulic pump; it doesn't require as large an engine as a single-stage model, because the pump supplies the force. Almost all, if not all, new splitters have this feature.

2.) Splitters are rated by the force they generate in tons: If you are serious about splitting your own fuel and plan to feed your splitter a steady diet of hardwood, you should buy a machine rated at 12 tons or more. A splitter dealer recently told me how he checks to see if a splitter has adequate power: He puts a 4- or 5-inch-diameter stick of hardwood on the tray sideways; if the machine can shear it in half, it passes muster. Wow!

3.) Hydraulic splitters are quite safe to use because the wedge (or plate) travels slowly. This is not true of the previously mentioned flywheel model, but no matter what kind you use, keep your hands clear of the wood, the shaft and the carriage when the pump is engaged.

4.) A splitter with controls and loading accessible from both sides will generally be more convenient to operate than one without.

5.) Make sure that the splitter can handle wood the length you use before you buy it.

6.) Most full-sized, well-made splitters have good-quality, highway tires. This is a desirable feature, but be aware that, since they have no springs or shock absorbers, splitters shouldn't be towed at speeds in excess of 30 mph. The bouncing doesn't do them any good, and they have been known to roll. A lot of people haul them on a trailer or in the back of a pickup truck.

7.) A splitter with loading trays keeps wood from rolling off of the carriage prematurely—a very worthwhile feature.

8.) Most splitters are made now with a brass liner for the carriage. This is a good feature because the liner can be more easily and inexpensively replaced than the carriage.

9.) Four-way wedges designed to split large pieces of wood into four pieces with one stroke sometimes push the wood off the carriage instead of splitting it and probably require more power to be effective than you otherwise need.

10.) There is nothing wrong with an electric splitter, so long as its power rating is adequate and you'll never need to use it more than a couple of hundred feet from an electrical outlet.

People who use hydraulic splitters frequently tell me that the most efficient way to work is with two people; one to operate the machine and the other to keep the machine supplied with wood and to keep the split wood out of the way. A little planning helps: Position the splitter as near the wood as possible and make sure that there is plenty of room for the end product.

The Stickler

This is a completely different sort of power splitting device: It looks like a much larger version of my lost and unlamented Wood Grenade, but it works. The power for the Stickler is supplied by your car. You block up the drive wheel, remove the tire and bolt on the Stickler, and you're in business. To operate it, put the car in low gear with the engine idling, and hold a piece of wood lengthwise—not by the end—against the tip of the Stickler. It will thread itself into the wood and split it.

This is quite an ingenious idea, and it has the advantage of being hundreds of dollars cheaper than a hydraulic splitter. It may have a few disadvantages, too. Most people don't hook up the emergency kill switch that attaches to the rear bumper of the car, within easy reach if a problem arises. A few people have subsequently had cause to wish that they had when a stubborn piece of wood has hit the ground just right while still attached to the stickler and thrown the car off of its jack stand. Also, the wood revolving on the stickler digs a hole in the ground, making a mess. Sticklers aren't as fast as hydraulic splitters, and finally, your car will spend a lot of time idling when you use your Stickler: You might have to add the cost of tuneup to the price of your wood.

Regardless of which tools and techniques you settle upon, splitting your own wood is time and energy-consuming labor. Another way to view the task, however, is to see it as one of the ways in which wood warms you: with satisfaction as well as with BTUs.

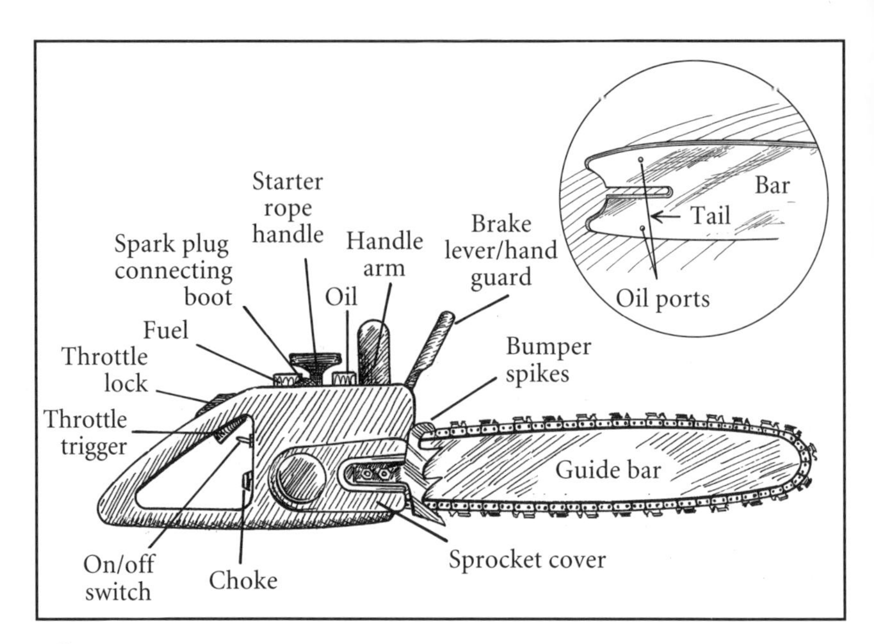

Chapter 5

Chain Saw Selection and Maintenance

The earliest chain saws may not have been true labor-saving devices. Introduced in the 1930s, the machines were heavy, relatively slow, vibrated with hand-numbing intensity and required two men to operate them. Time and evolution have been kind to the chain saw and to those who use them: The typical logger today uses a saw that has an engine the size of a small motorcycle's but weighs 20 pounds or less. Modern saws are also equipped with antivibration dampers, an effective front-mounted muffler and a significant safety advance in the form of a chain brake, which stops the chain in milliseconds in the event of a kick-back.

It is, of course, very possible to cut wood without a chain saw, but I am convinced that, for most firewood cutters, time's constraints mandate this power tool which is, in its present form, a labor-saving device that has become a near-necessity. This is not to say that using a chain saw is easy work, for it is not. It is to say that, with proper training and a good saw, you'll be able to cut far more wood in an hour than you will with a hand saw. If time is not of the essence for you, by all means consider a bow saw or two-man crosscut: These saws are much cheaper, much quieter and far safer to use than a chain saw—they have, in short, all the advantages that a maul has over a hydraulic splitter, save one: They are enough slower than their powered counterparts that they will transform cutting a winter's fuel wood supply from a chore into an occupation.

BUYING A SAW

What sort of chain saw do you need? This depends somewhat upon your physical condition and the wood you'll be cutting, but I feel safe in saying that most weekend firewood cutters won't need a big, pro-sized saw and that they should avoid the tiny, discount store specials. A chain saw with an engine in the 50-60 cc size range should provide plenty of power for firewood cutting without the weight and the price tag of the pro models.

Features you should insist upon are:

1.) Chain brake. As I mentioned, the chain brake stops the chain in the event of a kick-back, an unpleasant event we've discussed in previous chapters. Kick-backs have killed and injured many people, and while a chain brake won't prevent the accident, it will mean the difference between getting hit by a stationary chain and by a chain turning at thousands of rpms.

2.) Antivibe. It's hard to appreciate just how much difference antivibration cushions make until you've used a saw without them. Numb, tingling hands, increased operator fatigue and, for some people, "white fingers," more technically known as Reynaud's Syndrome, an often painful loss of circulation to the fingers, are the lot of those who don't insist upon this option for their saws. Antivibe systems are simplicity itself, consisting as they do of rubber dampers at the points of connection between the saw's handles and its engine. The dampers absorb the vibration with surprising effectiveness, as I can attest: After spending several years using saws without antivibe, I developed white fingers—a bad enough case so that several fingers on each hand would become numb and turn a ghastly white when I so much as gripped my truck's steering wheel on a cool summer morning. Several years after I switched to saws equipped with antivibe, the affliction went away almost entirely. Needless to say, I'll buy no chain saw without antivibe, and neither should you.

3.) Throttle interlock. A simple mechanism that prevents the saw's throttle from being opened unless the sawyer's hand is gripping the chain saw's rear handle. It should be obvious that no sensible person wants his or her saw's chain turning unexpectedly; the throttle interlock assures that, barring carburetion or idle-adjustment problems, it won't.

Another option to consider is bar length. Most saws in the power range I recommend will come equipped with a 16-inch bar and chain. Since a saw so equipped can cut trees that are nearly 32 inches in diameter, the 16-inch bar should be more than adequate for most people's needs. If you do anticipate that you'll be dealing with larger trees on a regular basis, a saw this size will easily handle an 18- or 20-inch bar.

I won't recommend a particular brand of saw, because I firmly believe that the most important choice to make when buying a saw is not the maker but the dealer. Chain saws are generally tough and reliable, but they do break down from time to time, and a dealer who stands behind his or her product and who has a skillful service shop is absolutely essential. All saws are good when they work—most saws, anyway—but a discount saw that you can't get serviced properly is no bargain.

The best way to choose a dealer is to ask people who make their livings with chain saws—loggers, tree surgeons and firewood dealers—whom they recommend.

A good chain saw is not cheap—as of this writing, you'll pay $400-$600 for a new saw suitable for serious home firewood production—and you may be sorely tempted to save several hundred bucks by buying a used one. Sometimes, if you know the person selling the saw and why he or she is selling it, and if you know enough about saws to tell what you're looking at (and listening to, since, like a used car, a used chain saw should at least start), a used saw can be a real bargain. Most of the time, however, I think your best bet is to spend the money for a new one: In addition to the saw, you will be getting a dealer to stand behind it, you'll have a warranty, and you won't be buying a problem that has exhausted someone else's patience. Besides, a good saw will almost always last long enough to more than amortize itself.

SAW MAINTENANCE

The best specific advice concerning the maintenance of your saw will come from the dealer and the owner's manual. My experience has been that good saws thrive on hard use and minimal maintenance, but that the maintenance, minimal though it may be, is critical if your saw is to be a pleasure rather than a frustration to use.

The most important things that you can do for your saw's engine are: 1.) Mix the gasoline and engine oil properly; 2.) Keep the air filter clean; and 3.) Keep the chain properly filed (more on this in awhile). The best way to assure that you are using properly mixed fuel is to buy your 2-cycle oil from your saw dealer and to follow the dealer's mixing instructions. Different oils have different concentrations: Some get mixed with gas in a 1:16 ratio, some in a 1:50 ratio, and some in various ratios in between, so be certain that you know you've got it right before you pour it in your saw's fuel tank. If you've used too much oil, it isn't that big a deal; the engine will perform sluggishly and will smoke a lot, but it won't be permanently harmed. Once you've cleaned the spark plug and filled up with the correct mixture, the saw should be as good as new. Not enough oil in the fuel mix will bring very different—and serious—results. If your mixture is consistently a little lean, the saw will probably run beautifully but will have its lifespan shortened. If your mix is very lean, the saw's engine will likely seize up—possibly forever.

Cleaning the air filter is a simple matter and will pay dividends in your saw's performance and, possibly, longevity. Removing the air filter usually involves loosening three bolts (see your owner's manual). I make it a practice to take mine out and tap it sharply to remove the accumulated sawdust every time I refuel. From time to time—how often depends upon how much you use the saw—it's a good idea to thoroughly wash the filter. You can do this in hot, soapy water or clear (not mixed with oil) gasoline. If you use soapy water, rinse the filter thoroughly in clear water and, in either case, let it dry completely before reinstalling it in your saw. Washing the filter removes more of the fine dust than does tapping and, more importantly, removes the oil that builds up in the mesh or felt of the filter and inhibits the flow of air to the carburetor. One note of caution: It is tempting, when your air filter looks like the bottom of a neglected canary's cage, to scrape it clean with a screwdriver or brush. Don't give in to temptation. Scraping the filter may remove some of the fuzzy little hairs which are there to catch fine dust and oil droplets.

Unless you are mechanically inclined and have the time, patience and tools required for extensive small engine repair, the only other engine maintenance you'll need to concern yourself with is the occasional cleaning and/or replacement of the saw's spark plug and the periodic tightening of every nut, bolt and screw you can find, with the exception of the carburetor adjustment and idle-adjustment screws. My saws have seldom needed more attention than this, and when they have, I've taken them to the dealer. If you are mechanically inclined and are also inclined to save the dealer's repair labor charge, you'd be well advised to take a course in small engine mechanics, if possible, and to get hold of a copy of *Barnacle Parp's Chain Saw Guide,* by Walter Hall (Rodale Press, 1978), an entertaining and excellent source of information about chain saws.

Maintenance of your saw's cutting assembly—the bar and chain—is an ongoing process, but simple enough once you learn how, and well worth the trouble. Maintaining the bar really is simple; just scrape the oily, gunky sawdust out of the groove every day you use the saw, taking care to clear the oil ports (see the drawing at the head of this chapter, inset). A knife, old hacksaw blade or special tool available from most saw dealers will make short work of the job. In addition, reinstall the bar upside down every time you clean it to assure that it wears evenly, and check the edges of the rails: If you feel that they are starting to develop a burr, file them smooth with a flat file. This will help keep the bar rails from chipping.

Filing

Maintaining your cutting chain is not quite as simple—until you've had a bit of practice, at any rate—but no other thing you do will have as much immediate effect on your saw's performance. A well-sharpened chain makes chain saw work almost easy; a dull chain makes it some of the most futile make-work in the post-industrial-revolutionary era. A dull chain causes increased wear on the bar and engine, since the sawyer will probably bear down in an effort to compensate for those dull teeth, and increased wear, tear and fatigue for the sawyer, as well. I hope this convinces you of the importance of learning to file your saw chain. If it doesn't, try cutting the same piece of wood, first with a dull chain, then again with a truly well-filed one. That should do the trick.

Some people never learn to file, but take their saws to a dealer or sharpening service periodically. The reason that this approach doesn't work is that chains get dull easily and frequently; sawing one muddy log will do it. Running the tip of the bar into the ground once will do it also. Don't think that you can avoid these mishaps, either. If you use your saw, you'll dull the chain. Even if you miraculously cut only wood, the dirt and grit on the bark and the wood fibers themselves will dull the chain, albeit slowly—so learn to file. The best way to do this is to have someone who knows how, such as your saw dealer, teach you. Diagrams and written instructions can help, but they won't get the job done. I do, however, have some general pointers that will apply to almost any chain you file:

1.) When you buy your saw, buy a supply of the proper-sized round files. (Your dealer will match the file to your chain.) Also buy a filing guide or jig (FIGURE 1) and learn to use it. Many—perhaps most—pros file freehand, and in time you may, too, but chains last longer and cut better when their tooth angles are always correct, and a guide assures you that they are.

2.) Hold the file firmly against the cutting edge of the chain tooth so that it applies even pressure along the whole edge (FIGURE 2). This

FIGURE 1.
Filing guide assures properly angled cutting edges.

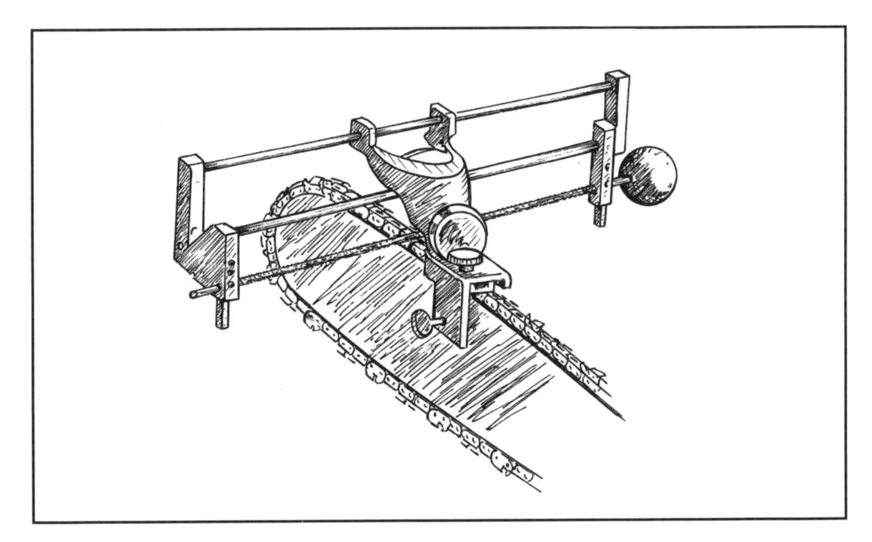

will help you avoid tipping the file one way or the other (FIGURES 3 AND 4) and putting too much or too little hook in the teeth. Too much hook shortens the life of the chain and causes unsmooth cutting, and too little hook leaves little or no cutting edge.

3.) Get a raker gauge and use it every other time you file to check your rakers (FIGURE 5). The rakers act as depth gauges; if they are too high, the sharpest chain in the world won't cut well. If they are too low, the teeth will constantly be biting off more wood than they can chew, and this will make the saw bog down or even kick back. File your rakers to the correct height with a small, fine, flat file. Be sure to retain the original profile of the rakers, as their shape is engineered to minimize the likelihood of kick-back.

4.) Every once in awhile, take the chain off and file the drive links with your round file (FIGURE 6). Doing this will help them clear debris from the bar groove. You may also have to work on the drive links—in this case with your flat file—if your chain comes off during use and they get damaged. You'll know if this has happened because the drive links won't fit back into the bar groove, or will do so only with difficulty. Just take the chain back off and file the burrs off of the drive links until they fit easily back into the bar groove.

5.) Don't hang onto dull files. Recycle them, make sculptures with them, but don't waste your time filing your chain with them. My logging partner had a strict rule: After four filings, he'd toss the old file and use a new one. I confess to peering into the bushes looking for his rejects from time to time, because the rule seemed overly strict, but the idea that inspired it is sound. Different files have different useful lifespans, but if you find yourself having to bear down to take metal off your chain, or if

FIGURE 2.
File on the cutting edge.

FIGURE 3.
Tipping the file one way gives the teeth too much hook. . .

FIGURE 4.
. . .tipping the other way gives them too little hook.

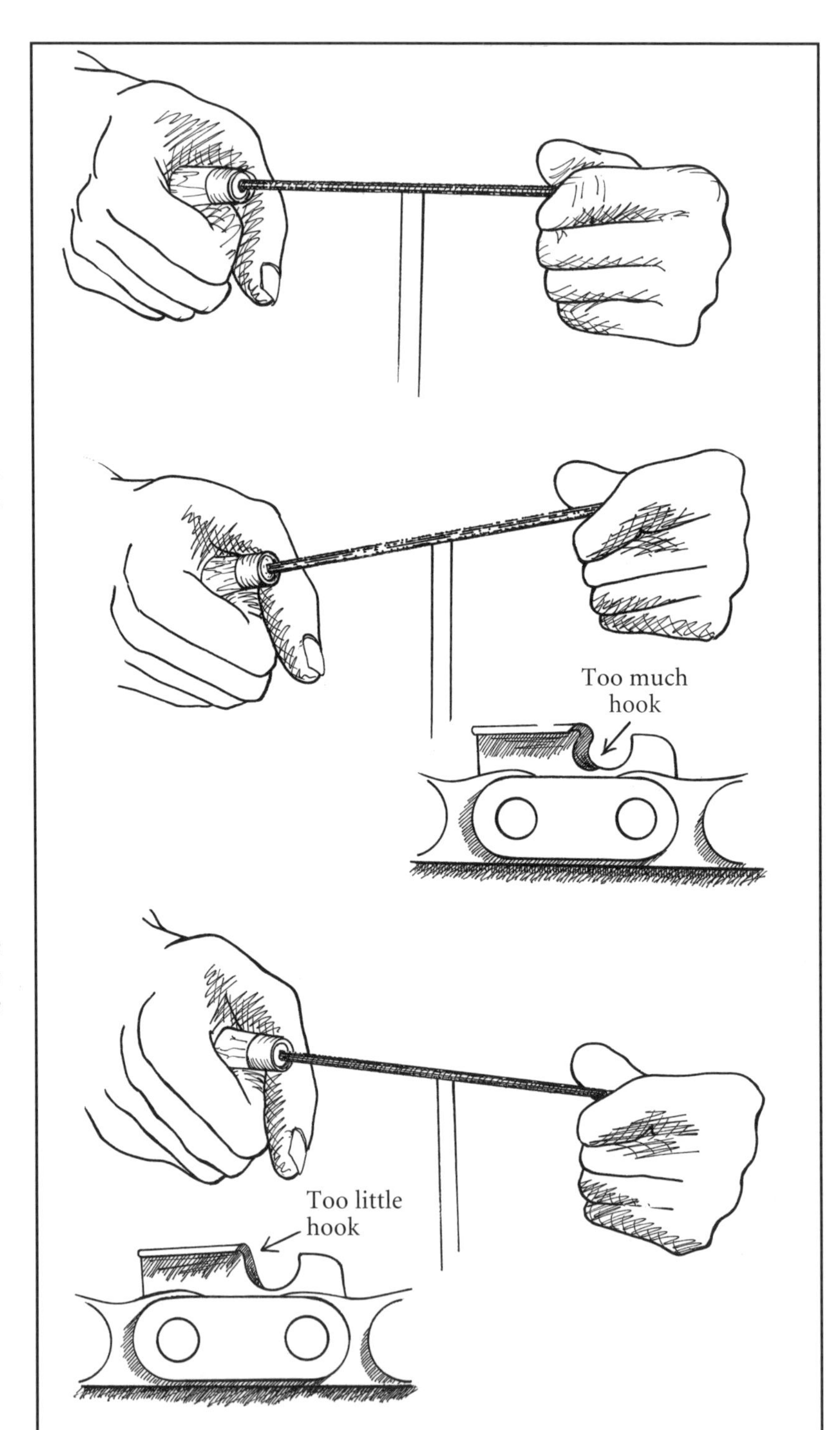

FIGURE 5. *Rakers must be filed to correct height.*

FIGURE 6. *Drive links*

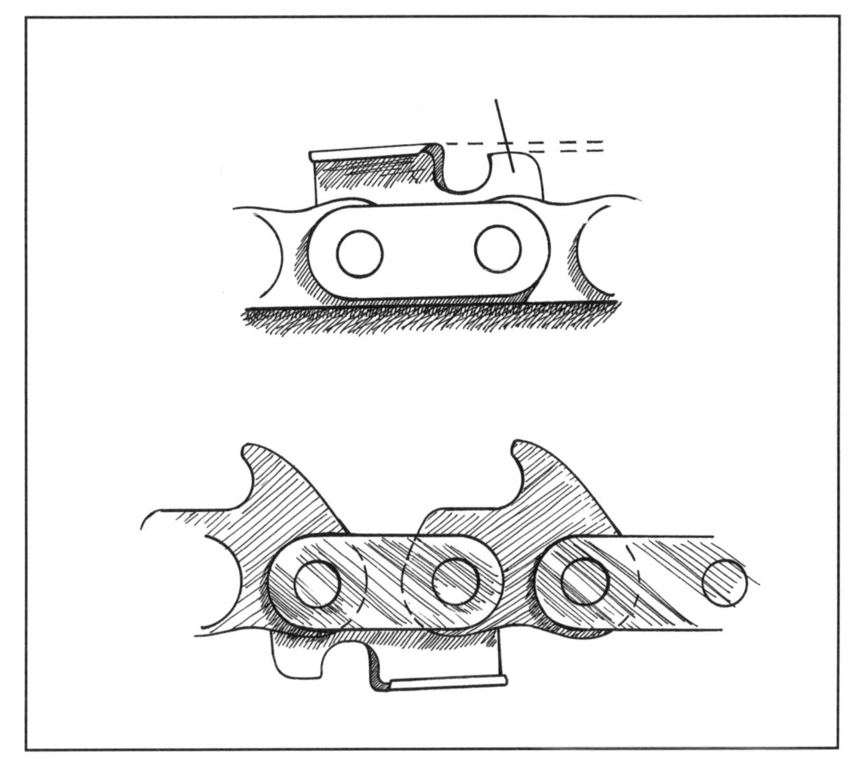

you notice any shininess on the file's cutting surface, discard it.

6.) Sharp chains—more accurately, sharp teeth—don't reflect sunlight.

7.) Well-filed chains don't throw dust; they throw chips.

8.) Make sure that the chain is properly tensioned on the bar before you sharpen it. A loose chain wobbles too much for accurate filing. Your chain tension is correct when the straps, or connectors, on a cold chain (a hot chain—one which has just been used—will be looser) just touch the bottom of the bar, and the chain turns freely.

9.) File only on the push stroke.

10.) In order to keep your saw cutting straight, it's important that the right- and left-hand teeth on the chain be the same length and be filed to the same angle—all the more reason to use a filing guide. If your chain takes a square hit from something unforgiving—a rock, a sugaring tap, a horseshoe—you'll have to look it over until you identify the most damaged tooth. File it first, then file all the others to its length.

If you take these tips to heart, get some good instruction and a lot of practice, you'll become an able filer. If you spend more than a few minutes a year using a chain saw, you'll find that the time you spent learning to file was time well spent indeed.

CHAIN SAW SAFETY

Over the last few years, chain saws have not only gotten lighter and faster, they have gotten safer as well. Chain brakes and improved chain designs have substantially reduced the risk of kick-back, but make no mistake: A chain saw is likely the most dangerous tool you will ever use. I know whereof I speak.

I was working fast one day a few springs ago, and in two hours the first two trees were down, cut up and piled. The third tree was the toughest, though, so I couldn't start resting on my laurels. It leaned heavily toward a shiny new rail fence—too heavily to be wedged—so I had to climb it and tie a pull-over rope three-quarters of the way to the top. Then I pulled the truck into the field, hooked a come-along to the rear bumper, and tied the pull-over rope to the end of the come-along cable so my helper could crank the tree over as I cut it. That went well; the tree fell as planned, and I was limbing it.

The operation was going nicely, and I was far enough down the tree so that the branches were 8 or 10 inches in diameter. The saw I was using was a very fast pro-sized saw—so fast that it would cut completely through a 10-inch-diameter stick of sugar maple before the stick began to fall. I was cutting left-handed (I'm right-handed) because the uncut portion of the tree was to my right and would have interfered with the saw had I done otherwise. The branch I was cutting was up in the air at about head level, and my modus operandi was to cut through, swing the saw in a semicircle to the next cut, and cut again: Fast, yes, but not very deliberate. Then one of those chunks of sugar maple that weren't falling until after the saw had passed through dropped squarely on top of the bar while the saw was swinging down. I had no time to react, and the weight of the wood combined with the downward momentum of the saw drove the bottom of the bar, with its rapidly turning chain, into my right leg just above the knee.

The friends I was working for that day, Jack and his wife Bobbie, were nearby. The cut was large, the entire width of my leg, and very deep. I discovered that it is unsettling to unexpectedly see your own femur and kneecap. I lay down, elevated the injured leg and held the cut closed while Bobbie, a member of the local rescue squad, went for tape and bandages and to call the ambulance. The presence of my quick-thinking friends emphasized an elementary but oft-neglected chain saw safety precaution: Never work alone.

As it turned out, I was lucky: After a bunch of stitches and some time on crutches, I was as good as new, the only permanent changes being a big scar and a different attitude.

At the time of this accident, I had relied upon good work habits, quickness and experience to keep me out of trouble for almost fifteen years as a professional sawyer; an earlier, minor cut and a few near-misses had not persuaded me to wear protective equipment, but this nearly permanently disabling accident, caused as it was by overconfidence and work habits that weren't as good as I had thought, was unsettling.

The second day of my recuperation, I ordered my first safety gear: Two pairs of pants with layered polyester pads.

I had used protective chaps years before while working for the United States Forest Service, but only when I had to. Those chaps, the forerunners of today's gear, were bulky, heavy and cumbersome. They certainly slowed me down, and that made me nervous. When you are cutting trees, you face two distinct hazards: the tree and the saw. Protection from one shouldn't make you more vulnerable to the other, and I was pleasantly surprised to find that modern chain saw safety clothing is a vast improvement over the bulky chaps I remembered. I don't feel comfortable working unprotected anymore and have, in fact, added a hardhat with hearing and eye protection, and safety boots to my wardrobe. To help you decide what to buy, let's review what's available.

Pants or Chaps?

Safety pants, which are available in a variety of styles, offer several advantages over chaps, the most notable of which is that when you wear them to work, you're protected. You won't take them off because they're too warm or because you aren't using the saw at the moment—unless, of course, immodesty and the law hold no terrors for you. It's too easy, with chaps, to stow them in the truck when you think you are done and not go to the trouble of putting them back on for a few more cuts.

The other major advantage of safety pants is that they usually protect more of the leg than chaps do. Also, because they don't have straps, they are less likely than chaps to get snagged by brush.

The disadvantages of safety pants include their price—$70 or more compared to $50-$60 for chaps—and their considerable warmth, fine in January, but uncomfortable in August, even the summer-weight models. I was tempted a few times, on very hot days, to do without the protective pants, until one of my daughters pointedly remarked that I must have enjoyed the air conditioning in the hospital emergency room the day of my accident. I have since bought chaps and have been very good about wearing them when I'm cutting; all protective garments reduce the risk of injury, but only if you use them. They don't help at all when they're hanging in your closet.

The Protective Materials

Ballistic nylon has been the standard component of the protective pads in both safety trousers and chaps for years, but it is rapidly being replaced by Kevlar (a registered trademark of DuPont). Some tests have indicated that Kevlar provides better protection with less weight and bulk than nylon, but the more there is of either, the more protection it provides. Both Kevlar and nylon act as barriers. They provide more resistance to a turning saw chain than a pair of blue jeans will, but they will not stop it; a typical safety garment with Kevlar or nylon pads will give you less than a second to react before the chain cuts your leg.

The Swedes, longtime leaders in devising safety equipment for woods workers, have a different theory: By layering loosely woven polyester, they get a pad that fluffs when hit by a moving chain and almost instantly clogs the chain and sprocket, stopping it cold. No manufacturer will guarantee that these pads can't be cut through, but in test after test there have been very few failures, even with full-sized professional saws. A chain saw dealer of my acquaintance ran such a test as a promotion to introduce the complete line of safety equipment he was selling. He invited area loggers to cut through a piece of the polyester pad with their own saws, and chuckled a little as he recounted that none of them could do it.

What to Buy

The choice involves the style of the garment (pants or chaps) and the composition of the protective material (ballistic nylon, Kevlar, or loose-weave polyester). As I mentioned before, the pants provide more protection and are perfect for sawyers who usually do their chain saw work in colder weather. If you do your cutting in all kinds of weather, or intermittently, you may prefer chaps. As for the material, the polyester pads seem to offer the most protection, and not just in controlled experiments. I ran into two friends, both woodsmen, at the local general store awhile back. One of them, fresh from work, was wearing pants with polyester pads, and they had a big gash above the knee. "I tripped," he said, "and the saw went right into the knee and stopped. The chain was really plugged, but I don't have a bruise." The other guy told about his accident, quite similar to mine. He got cut, though not very seriously, right through his Kevlar chaps.

A note of caution before you buy padded pants or chaps: Some tests have indicated that the polyester pads, while extremely effective at plugging saws with closed-rim sprockets, are less effective at stopping saws with star sprockets. Most large saws have closed-rim

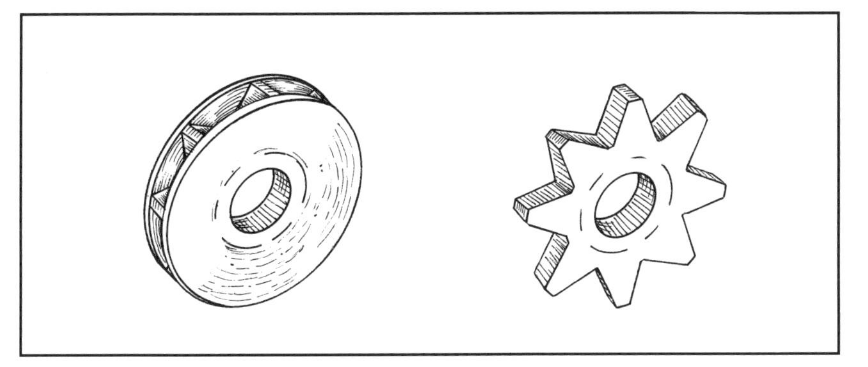

FIGURE 7. *Closed rim and star sprockets*

sprockets, and many midsized models do, too, but check to be sure. To do this, remove your saw's sprocket cover. The sprocket is the small wheel that the chain loops around; compare yours to the those in FIGURE 7. If you still aren't certain which kind of sprocket you have, ask your dealer. It is sometimes possible to retrofit a saw with a closed-rim sprocket if it came with a star.

One other note about safety pants and chaps: If you do cut them, discard them unless you can replace the damaged pad. It's tempting to perform a superficial repair with needle and thread rather than throw away an $80 pair of pants, but it's a mistake: Once a pad has been cut, it offers little or no protection. A trip to the emergency room will cost far more than a new pair of trousers.

Other Safety Gear

As I and many others have noted, woodcutters tend to be a macho and fatalistic breed, slow to accept personal safety gear and adept at rationalizing their reluctance. To illustrate, after years of bareheaded timber felling, I started wearing a hardhat with integral eye and ear protection. As luck would have it, on my first day with the hat a limb nailed me on the top of the head, leaving me with ringing ears and a sore neck, but without the concussion or fracture that would have been my lot the day before. "It wouldn't have hit you if you hadn't been wearing that hat," said a co-worker, smiling since he knew what nonsense that really was. Still, it was more than two years before he got a hat of his own.

Often, as in my case, it takes an accident or a frightening near-miss to stimulate a woods worker to get protection. But, you say, of course loggers should use safety gear; they use saws all day, every day. What if I only cut firewood a few days a year? Do I really need a pair of expensive safety pants and all the other paraphernalia that goes with them? The answer is that it takes just a few hundredths of a second for a chain saw to disable or kill you, and it is the nature of probability that your exposure to the danger is as great during one day as is a logger's during one day. Perhaps greater, given that you likely are less experienced and skilled with a saw than are most loggers.

In addition to safety trousers or chaps, anyone who operates a chain saw—infrequently or not—should have head, ear and eye protection. Head protection is most crucial in the woods, but even if you are bucking wood in your back yard, a hardhat provides a measure of protection against kick-back, flying wood and broken chains. The need for ear protection will be obvious to anyone who hasn't already damaged his or her hearing by running a saw without it; even the new models with their improved mufflers will unquestionably damage unprotected ears, temporarily and permanently. Eye protection not only fends off those seagull-sized wood chips that your saw throws at you, it makes you a better sawyer; because you can see your work, your cuts are more accurate than they are when you are flinching and blinking.

You can get head, ear and eye protection separately and quite cheaply: A hardhat, a pair of safety goggles and some ear plugs probably won't set you back $20. The other choice is the integral head protection system I mentioned earlier. At $30-$40, it is a good deal more expensive, but it has some advantages which, in my estimation, make it worth the extra money. In the first place, the protection is together; you won't lose your ear plugs or glasses, because they're attached to the hat. In the second place, the eye protection is fine-mesh screen (you can flip it up out of the way when you don't need it); it doesn't fog up the way goggles do, and protects your nose as well. In the third place, these logger's helmets are comfortable.

People do cut their feet with chain saws, and they also have heavy pieces of wood fall on their toes, so it follows that a pair of steel-toed safety boots should be part of your sawyer's uniform. Several manufacturers make safety boots specifically for saw operators: These boots have, in addition to steel toes, layers of Kevlar through the whole boot. They come in summer and winter weights, are comfortable and well made, and in my experience, justify their considerable price (over $100) by outlasting ordinary workboots nearly two to one.

Hand protection is another consideration. Leather gloves not only protect from cuts and scratches, they give better purchase on the saw and cushion vibrations that can cause fatigue. Gloves and mittens with protective pads in the backs are available and provide an extra measure of protection, since your front hand is likely to be knocked off of the saw's handle in the event of a violent kick-back.

I've recently seen safety shirts on the market. These have protective pads in the shoulder and chest area, and while I don't worry very much about getting cut there, because of my chain brake and somewhat discredited work habits, I can't help but remember that I never figured I'd cut my leg, either.

If I had been wearing safety pants or chaps, I probably wouldn't have been cut. Some readers will note that I wouldn't have been cut had I been working safely, with or without protection—and that is probably true, but I know very few professional sawyers who have not cut themselves, and I have to conclude that to err is human, after all. To err with a chain saw can have such awful results that it only makes sense to get as much protection as you can.

Chapter 6

Your Chimney

We've been talking about cutting wood for awhile; now it's time to consider burning it. If you are not currently a woodburner but are considering becoming one, you have quite a few choices and decisions ahead of you. If you already use wood, it won't hurt to consider your wood heating system: Perhaps your enjoyment of it could be enhanced, as well as its safety and efficiency.

I cannot overly stress that a wood heating system is just that—a system. The same stove may not perform the same in two different chimneys; the same fireplace may smoke and smolder in one house but not in another; the same stove and chimney in the same house may behave differently for two different operators. It is essential, if woodburning is to provide the warmth and pleasure of which it is capable, that all the elements of the system—the chimney, the house, the appliance and the operator—work harmoniously together.

NEW CONSTRUCTION

We'll begin by addressing those who are planning to build a new chimney for woodburning, whether as part of a new house or an existing house. As a chimney sweep, I frequently come across people who have had houses built with fireplaces or woodstove hookups who have problems—major and minor—with their heating systems. The problems, I believe, stem from the process: Typically, the only decisions the homeowner makes are along the lines of, "I want a woodstove in the living room," and later on, picking a color for the hearth brick. These decisions are rarely made based upon an understanding of their future implications, and neither, unfortunately, are subsequent decisions concerning such critical matters as the specific location of the chimney and its flue size, even though these decisions are usually made by architects or contractors. This process does not consider the wood heating system in any unified way and can result in houses with expensive and attractive but unusable wood heating components. This is a pity, because, particularly in the case of a new house, good planning can make wood heat easy to live with.

I recently saw a case in which a family had a beautiful house built with a chimney in the living room for a woodstove. They moved in, bought and installed a stove and, when the weather got cold, fired it up. What they got, along with their heat, was a large amount of liquid creosote dripping on the hearth every time they started the stove and an alarming buildup of flaky and glazed creosote in the chimney. There were two—possibly three—reasons for the problem.

In the first place, the chimney was located on an exterior wall, thus assuring that it would rapidly cool the flue gases produced by the stove, resulting in the condensation and deposit of tars and creosote. Why was the chimney on the outside wall? Because that was where the architect's drawings placed it, probably to save space.

The second factor was the flue size—too large for a high-tech stove, which requires a chimney with draft velocity rather than volume. This was a decision made by the masonry contractor who, not knowing what kind of stove was going to be used, built a chimney that could accommodate—he thought—any stove.

The third factor—and it may not prove to be a problem after all—was the stove itself, which may be too large for the very tightly constructed and well-insulated house; if the stove makes the room too warm when it is operated properly, it will probably be damped down too far and too often for clean burning.

In this case the solution to the creosote problem was fairly easy, though expensive: The chimney was relined with a smaller stainless steel liner and insulated. It may ultimately be necessary to replace the stove as well, but for now the system seems to be working quite well.

Almost any chimney sweep you ask will have had numerous similar experiences. The other common—and frustrating because they are avoidable—problems I often encounter in new construction are chimneys that do not allow a stove and/or smoke pipe to be installed with the proper clearances to combustibles, and fireplaces that smoke because no provision was made to give them combustion air.

You can avoid costly, infuriating and dangerous problems with wood heating systems in new construction if you know enough to design the system as a whole. To a degree, the house needs to be designed around the system, rather than vice versa. I strongly suggest that you hire a qualified chimney sweep or other wood heating specialist as a short-term consultant: The money you pay him or her will be a drop in the bucket compared to what it would probably cost to fix a system that doesn't work properly.

In order to help you with the decisions you'll need to make when you're choosing a wood heating system, here are some general bits of information:

1.) Chimneys that are in the house for most of their length stay cleaner, work better and return more heat to the house than do chimneys outside the exterior walls.

2.) Woodstoves should be sized to the space they must heat. A stove too large for its space will provoke you to burn it slowly, causing creosote buildup. A stove that is too small may lead you to overfire it, possibly damaging the stove, the chimney and, should either fail, the whole house.

3.) The chimney flue should be sized to the appliance it serves. Ideally, if your stove has a 6-inch round outlet, your chimney's flue will also be 6 inches in diameter. A larger flue may work well, but at a certain point (depending upon the height of the chimney, the stove itself, and other factors specific to each situation), a flue that is much too large for a stove will result in problems such as sluggish performance and rapid creosote buildup. A flue too small for a stove, though unusual, is unacceptable also, and will have similar symptoms. Fireplaces do sometimes have undersized flues. Since it relies upon a large volume of air to maintain its draft, a fireplace with an undersized flue will admit more air than the chimney can expel, with the result that smoke will back up into the house.

The area of a square or rectangular chimney flue is not as easy to calculate as a person with my primitive math skills might assume: a nominal 12-by-8-inch flue does *not* have an area of 96 square inches, in the first place because its inside dimensions are probably really only 10½ by 6½ inches, and, in the second place, because the corners of the flue are rounded. The flue's *actual* area is a good deal closer to 70 square inches. FIGURE 1 gives the formulas for determining the areas of square and rectangular flues. (A round flue's area is determined by multiplying its radius squared by 3.14, or *pi.*).

Any competent mason will know the proper size relationship between the flue and the fireplace. (If you are interested, the area of the flue should be approximately one-tenth that of the fireplace opening. A short chimney may require a larger flue.)

4.) Fireplaces and, sometimes, stoves may need cold air returns to work properly in tightly constructed houses.

5.) Chimneys must be accessible for cleaning from both the top and the base. If access is difficult, it may be only your chimney sweep's problem (though it could cost you more). If access is impossible, it's your headache.

6.) Solid concrete blocks are an acceptable chimney construction material, but shouldn't be used as the exterior layer where they'll be exposed to the elements, because they can deteriorate rapidly, more than offsetting their low initial cost.

7.) Each appliance should have a separate flue. This is a provision of the National Fire Protection Association code, but is widely disregarded, since multiflue chimneys are much more expensive than single-flue chimneys. The risks of sharing flues include inadequate draft and the possible venting of toxic gases into the living space. Also, if both your wood heater and furnace use the same flue, a flue blockage will leave you without a heat source until the problem is rectified.

8.) Chimneys and thimbles (the opening in the chimney for the stovepipe) must be located so that the appliance can be installed in accordance with clearance requirements. It's shockingly common to find, even in new houses, thimbles 4 or 5 inches from a combustible ceiling; they should be at least 18 inches below it. (More on clearances later.)

9.) The less stovepipe an installation requires, the better. Smoke cools rapidly in the pipe, leading to creosote formation. Having less pipe will allow you to run the stove hotter since you'll get less heat from the pipe.

10.) Factory-built, insulated metal chimneys that are listed by Underwriters' Laboratories for venting wood heating appliances are just dandy. They've gotten a bad reputation in some quarters due to past failures in chimney fires, but the new models stand up to intense heat as well as, and usually better than, standard masonry chimneys, and most sweeps will agree that they tend to stay cleaner and are very well matched to efficient modern stoves.

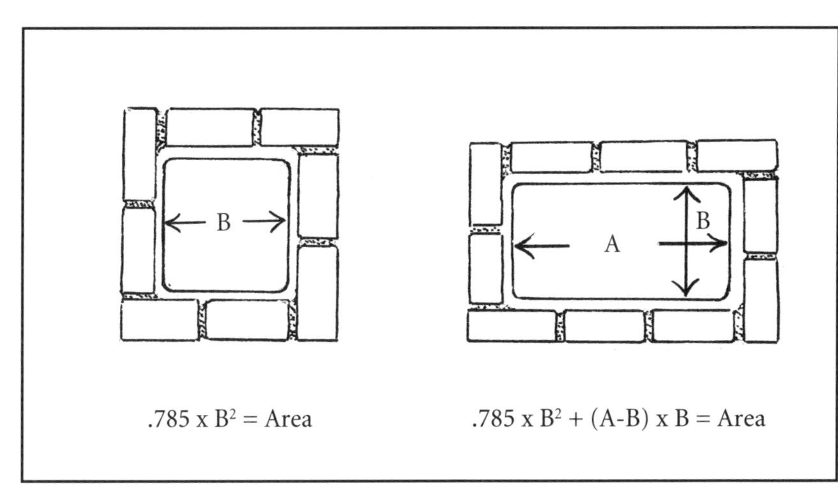

FIGURE 1. ***Calculating areas of square and rectangular flues***

EXISTING CHIMNEYS OF RECENT VINTAGE

This section is for people with fairly new and apparently sound chimneys. It's quite a temptation, as we enter the age of high-cost energy, to plug a woodstove into your chimney and enjoy cheap heat again. Before you run out and buy a stove, though, you need to spend some time evaluating your chimney. Again, I suggest that you hire a qualified sweep to inspect it and make recommendations, not because I'm trying to drum up business, but because I've come to realize that a high percentage of existing chimneys, unless they are modified, are unsuitable for wood heat. A good sweep can save you from having a lot of serious problems. Here are some of the factors he or she will consider:

1.) Since the chimney is fairly new, it's probably lined with clay tile, but is the lining sound? Are there cracks, loose mortared joints or other deterioration?

2.) If your chimney is factory-built metal, is it sound? No buckled seams? Is it appropriate for woodburning? Were required clearances adhered to in its construction? Were all of the necessary support and shielding accessories used?

3.) Will the chimney provide a flue solely for the use of the woodstove (see previous section)?

4.) Will the location of the chimney and thimble allow for a safe installation—one that abides by required clearance standards? If not, can the problem be rectified?

5.) Are there inherent problems, such as flue size and chimney location, that may affect your choice of stoves?

6.) If the chimney and stove are on opposite sides of a combustible wall, does the wall pass-through provide for adequate clearance? Don't be lulled into a false sense of security by brick or stone facades; be certain that the connection is safe.

7.) If you are considering activating an unused fireplace, you need to evaluate the condition of the flue, damper, firebox and smoke chamber, as well as the chimney. Was construction done according to applicable codes? Is the system free of leaves, birds' nests and other debris?

OLD CHIMNEYS

You own a lovely old farmhouse in the country. The views are magnificent, the sills and beams are fairly sound—and you spent $3,000 on heating oil your first winter there. Come spring, you hop into your truck, drive to the nearest stove store, and buy the latest, most efficient woodstove your dwindling cash reserves can afford. You bring it home along with a few sections of stovepipe and a fire-proof hearth mat, call a couple of rugged friends, and install the stove in the handsome old brick kitchen chimney. Now all you need are a few cords of wood, and your new purchase can begin to amortize itself.

Then it occurs to you that the chimney probably is dirty, so you call a chimney sweep, who's glad to check your chimney. He arrives, sets up ladders, sweeps, then gives you—along with his bill—some bad news: Your chimney isn't lined. It is one layer of bricks held together by crumbling hundred-year-old mortar, its own weight, and inertia. The sweep explains that the house and chimney were built long before it was common practice to line each flue with clay tiles. Furthermore, your chimney can't be used safely to vent your stove until it is lined; excessive heat, sparks and even flames can escape from your chimney through any of the weak spots in the mortar and kindle a fire in areas of your house other than the stove's firebox.

What's more, the chimney sweep continues, an unlined chimney can't be cleaned as well as a lined chimney; he can't reach with his brush the nooks and crannies between bricks where creosote collects.

After the sweep leaves, you ponder the problem. Maybe he's being an alarmist: People burned wood for hundreds of years without lined chimneys. But old-timers didn't have airtight stoves. Airtights can cause a greater and more rapid buildup of creosote than is common with leaky old-fashioned stoves, and even clean-burning catalytic stoves may perform poorly in old chimneys, since the flues are liable to be too large, and the rough, unlined interior will provide too much resistance to the smoke. Also, you've noticed a few burned-out cellar holes here in wood heating country. You envision your house as a charred ruin and decide to have the chimney lined.

How can you tell on your own whether your chimney is lined? If the house was built prior to the 1940s, there's a good chance that the chimney is not lined. Take a look for yourself; if the chimney isn't too dirty, you'll probably be able to tell immediately whether it's lined. Look down the chimney with a good flashlight, or look up from the cleanout door or thimble using a mirror (a bright, sunny day when the sun is high is the best time for this). If the chimney is unlined, the inside will look like a brick wall. If it's lined, you'll see smooth tiles, not bricks, and the joints between tiles will run horizontally. If you aren't sure what you're seeing, it's time to call a chimney sweep for an inspection.

The presence of tile liners does not automatically mean that the chimney is completely lined; some old chimneys have been rebuilt from the roof line up with tiles but remain unlined from the roof down through the house. Nor should you assume that a tile-lined chimney is in good enough repair to be safe; time, chimney fires and the normal temperature swings associated with wood heating can cause the mortar between the liner's joints to deteriorate and fall out, leaving gaps in the protection. Water, earthquakes and normal settling can also render a chimney that appears sound unsafe. I tend to trust recently built tile-lined chimneys, but I've learned that with chimneys, as with most things, looks can be deceiving. Thoroughly inspecting a chimney before using it and periodically thereafter is a sensible precaution to take.

On request, a chimney sweep can inspect a chimney after he cleans it by lowering a strong light through it, if the chimney is straight and not too tall, or by sealing the chimney and inserting a smoke bomb. Either method will probably reveal major damage if it's present, and the light may show some incipient problems, but a truly definitive inspection probably requires the use of a miniature video camera specially adapted for the job. This is an expensive procedure—usually several hundred dollars—and, given the high cost of the equipment involved, many sweeps don't do it, but it's probably worth doing if there is any reason to suspect the chimney's soundness: A chimney that fails at an inopportune time may well cost you far more than would a thorough inspection.

Lining Options

In an age that has produced such technological marvels as microwave ovens and inflatable basketball shoes, it may come as a surprise to some readers that only three materials are used for lining chimneys: thermal concrete, stainless steel, and clay tile. Each method has advantages and disadvantages, and there are subgroups within the three major groups.

Thermal Concrete

The process of installing a concrete chimney liner usually—but not always—involves placing an inflatable bladder in the chimney and pouring a special concrete mixture into the chimney around it. Once the concrete has set, the bladder is deflated and removed and—presto—the chimney is lined. Most of the companies offering concrete lining use this method, but at least one—the Ahrens Company—fills the chimney with no-slump concrete (concrete with a very low moisture content that stays where it's poured) and then pulls a vibrating bell up through to form the flue.

The concrete used for chimney lining differs from the material used to make your basement in this respect: Instead of mixing sand and gravel with water and cement, chimney lining companies use lightweight, granular insulating material such as perlite in place of the sand and gravel. The result is concrete with very good insulating quality (good for maintaining warm flue temperatures) that is relatively light and so is less likely to induce the collapse of an old chimney.

The advantages of thermal concrete as a chimney lining material include: 1.) Its insulating qualities; a warm flue tends to stay cleaner and is less subject to damaging condensation than a cold flue. 2.) The liner is smooth and seamless, thus providing very little obstruction to smoke traveling through the chimney and offering no joints that could separate and become gaps. 3.) Thermal concrete has no esthetic impact; it doesn't change the outward appearance of a chimney. 4.) Long-term warranty—as much as 20 years—contingent upon proper use and maintenance.

The disadvantages of thermal concrete include price. This is probably the most expensive chimney lining technique, rarely costing less than $1,000 and often a good deal more. It is also out of the question for a do-it-yourselfer seeking to save some money: The equipment is expensive and generally available only to franchised dealers.

Another drawback to thermal concrete flues is that cleaning them can be, though it rarely is, a problem. The material is fairly soft, so standard wire chimney brushes can't be used, as they are too abrasive. The polypropylene brushes recommended for thermal concrete flues will usually clean them adequately, but probably would not remove heavily crusted or glazed creosote. The Ahrens Company avoids this problem by making a second pouring of a hard refractory material; the thin inner layer that results is abrasion-resistant. In any case, thermal concrete liners, because they tend to stay warm, are not very likely to house hard creosote formations.

Finally, although this is not a disadvantage so much as it is a cautionary note, any chimney lining technique, including thermal concrete, can be difficult to apply to a flue that is not straight. Offset flues are common in old chimneys, and since it is critical that the flue liner be surrounded by adequate air space or insulating material if dangerous exterior hot spots are to be avoided, an acceptable job must include the installation of spacers to keep the flue centered. This may involve breaking into the chimney at the offset, if feasible, or careful measuring and installation of the spacers on the bladder. Either is a pain in the neck for the contractor but will be done anyway if he or she is conscientious. If your chimney has an offset, it wouldn't hurt to discuss with your contractor how it will be dealt with before the work begins.

Stainless Steel

There are two basic types of stainless steel chimney liners: rigid—stovepipe sections made of stainless steel and secured to each other by pop rivets—and flexible. Either can be installed by a consumer who has no fear of heights, a fair amount of strength, a little ingenuity, a lot of confidence and at least one willing helper with the same attributes. The idea is to put the liner in the center of the chimney, fill the space around the outside of the pipe with granular, nonflammable insulation such as perlite or vermiculite, and then rain-cap the pipe (to keep out water which will speed corrosion) and flash the space at the top of the chimney between the pipe and the outside edge of the old chimney to keep the insulation from becoming waterlogged. This relatively simple procedure becomes more complicated when the flue being lined is offset (see discussion of thermal concrete) and it becomes necessary to install elbows or spacers or both. Another hitch can occur if the flue being lined already has a tile liner and the reline is necessary because of damage to the tiles or to reduce the flue size. In this case, it may be necessary to remove the tiles in order to make room for the steel liner and the insulation, a job requiring specialized equipment.

In recent years there have been several refinements in both the products and procedures used in stainless steel lining: Hardware is now available for suspending the liner in the flue, and for sealing the spaces between the liner and the old chimney walls at both the top and bottom. Also, it is common and recommended practice now to mix a small amount of cement and water with the insulation in order to keep it from settling. There are, better yet, ceramic wool insulating blankets available which wrap around the liner and replace the loose fill insulation, and which will also make the use of spacers in an offset chimney unnecessary.

The question whether to use rigid or flexible pipe will be answered by your chimney and your bank account: If your chimney is straight and your wallet is thin, rigid pipe will be your best bet, since it is substantially cheaper than flexible. Offset flues make flexible look more attractive, since it can often be installed without breaking into the chimney, but no matter what the state of your chimney and your finances, get several estimates; since most chimney sweeps install stainless liners, there is likely to be a fair range of prices available. Be sure to discuss with each bidder what the price includes: How will the flue be insulated? How will he or she deal with offsets (if any)? What does the manufacturer's warranty cover, and what will the contractor guarantee? Does the contractor understand that the liner is subject to thermal expansion, so the anchoring and weather sealing at the top of the chimney must allow for a good deal of natural movement? Is the liner material solid stainless? It should be, and will probably be type 304 stainless, which has acceptable resistance to both heat and corrosion. If another type is suggested, find out why and get a second opinion, since different types of stainless steel possess different qualities of heat- and corrosion-resistance. Also, if the liner is to be rigid pipe, it should be 24-gauge or thicker (22-gauge is thicker, 26-gauge is thinner). Flexible must be lighter gauge to be flexible.

The advantages of stainless steel chimney liners include: 1.) It is possible, in some cases, for a homeowner to install a stainless liner, thereby saving hundreds of dollars. 2.) The finished product is a safe liner, warm and securely connected. 3.) Even professionally installed, a stainless liner is usually cheaper than thermal concrete. 4.) It is comparatively easy to clean.

The disadvantages of stainless are that although it may be cheaper than concrete, it still isn't cheap, and nothing lasts forever, including stainless steel chimney liners. The manufacturer's warranty, which usually covers the materials contingent upon correct installation and maintenance, will probably be for 10 years. In fact, a stainless steel liner may and probably will last much longer—20 years or more—if you take care of it. Conversely, you can ruin it by neglecting it: The inside of a chimney is a very corrosive place, and even stainless steel will deteriorate if it's never cleaned. (This, by the way, is why lining a chimney with standard galvanized pipe is, though inexpensive, a very bad idea, since it will probably fall apart within a year.) Also, stainless steel may lose some of its corrosion resistance if it is exposed to high-temperature chimney fires.

Clay Tile

It is sometimes possible to find a mason with the equipment and knowhow to line a standing chimney with flue tiles. This is a tricky proposition, because the only way to seal the joints between tiles is to apply mortar (it should be refractory cement, actually) to each tile before lowering it, and then line it up perfectly with the section of tile that preceded it. This can be done with the right equipment, though even then you can't be sure of the integrity of each joint. Done with the wrong equipment, this chimney lining technique is a complete waste of time and money. A case I encountered a few years ago perfectly illustrates my point: The contractor, lacking the inclination to do an honest job, lowered each flue tile with two pieces of piano wire which he ran under the tile in an X-pattern. Of course, he couldn't remove the wire once the tiles were in place, so he left it, resulting in a flue with two wires crisscrossing its interior at each joint; impossible to clean properly.

Why should you worry about a few gaps in a chimney liner, anyway? The answer is that those gaps can become large holes in a chimney fire. Also, creosote can work its way through an unsound joint into the space between the liner and the chimney wall. If it ignites—which it can—it is very difficult to extinguish and can quickly spread to the house.

The main advantage—indeed, the only advantage—of clay tile relining is that it should be substantially cheaper than thermal concrete or stainless steel and, done properly, is much better than no liner at all.

The disadvantages include: 1.) The aforementioned impossibility of assuring that each joint is sound; 2.) It is difficult to maintain the necessary ½- to 1-inch air space between the tiles and the inside of the chimney shell when relining; 3.) Clay tile is more likely to sustain damage in a chimney fire than stainless steel or—probably—thermal concrete; 4.) It may be difficult to find a qualified mason in your area who offers this service.

When Not to Line

Every chimney used with a woodburning appliance should have a liner, but this doesn't necessarily mean that lining your existing chimney is your best option. Sometimes it is cheaper or more sensible for other reasons to build a new chimney. Some old chimneys are so badly deteriorated that they would have to be rebuilt before they could be lined. Or an old chimney could be unsafe to use for reasons a liner will not correct, such as one I saw recently that had a floor joist running through the middle of the flue. Or perhaps the chimney will need to serve two or more appliances and will therefore need multiple flues. It isn't always possible to put two flues, let alone three, in an existing chimney, and even when it is, it may cost more than rebuilding.

CHIMNEY CLEANING

Burning wood in a stove or fireplace produces good things, such as warmth and esthetic enjoyment. But since wood fires don't result in complete combustion, burning wood also produces an undesirable product: creosote, a highly combustible mixture of tar droplets, vapors and other organic compounds. Creosote will appear in your chimney in different guises—from flaky and powdery to hard and glazed or even liquid. Which type of creosote you see in your heating system will depend on the system itself and on your habits of operation. Is your chimney exterior, and do you habitually operate your stove at a low temperature to avoid overheating your house? If so, the odds are good that you will find glazed creosote in your flue, since you regularly create conditions ideal for its formation: low-temperature fires that produce lots of smoke and don't result in enough draft velocity to move the smoke out of the chimney rapidly.

The opposite extreme is an interior chimney serving an open fireplace in which hot, active fires are burned. The warm flue, unrestricted oxygen supply and strong draft normal to this type of system would probably allow very little creosote to form, and that which did would be flaky or powdery. Between the two extremes are many possible variations: flaky creosote near the bottom of the chimney and no creosote at all near the top; no creosote at the bottom and glaze at the top. The list could go on, and it is this variety which makes a chimney sweep's life so entertaining, if you'll permit me a little joke.

Removing the creosote from a wood heating system is one of the ways in which woodburners pay the piper: It isn't fun, but it's absolutely necessary if the music is to continue. Creosote is bad for your chimney in three principal ways: First and foremost, it can ignite and burn at temperatures in excess of 2,000 degrees Fahrenheit. This is a chimney fire, folks, and while a number of veteran woodburners start them intentionally to clean their chimneys, the truth is that every time your house survives one intact, you've dodged a bullet.

Chimney fires resulted in $63.3 million worth of property damage in the U.S. in 1989. In some instances, the damage was minor, but in some instances lives and entire houses were lost. You may not even be aware of the harm a chimney fire has done, since it is often confined to the chimney itself; I'd have to guess that over a third of the chimneys I see after a chimney fire have sustained obvious damage to their flue liners, and I tend to suspect that no chimney emerges from a severe fire completely unscathed.

A case that sticks in my mind is one I encountered a few years ago. The house, perhaps ten years old, was owned by a family that used it for vacations. I was called to do a routine chimney cleaning by the property management firm that took care of the house in the owner's absence. I checked inside first, and determined that there was a fireplace and a woodstove that shared the flue. After assuring myself that dust and soot would stay in the chimney when I swept, I got on the roof. There I discovered that the chimney had suffered a fire of such severity that large pieces of the clay tile liner had broken away, leaving the chimney completely unsafe to use. Not only had the owners not realized that the chimney was damaged, they hadn't even known that they had had a chimney fire, possibly because it occurred while they were out for the day.

The second way in which creosote harms a chimney is less spectacular than a chimney fire, but a cause for concern, nonetheless: Due to its acidic nature, creosote left in a heating system will cause deterioration of mortar and metal components.

The third harmful effect of creosote is that it can completely block a chimney, due either to the compound's tendency to expand in a chimney fire, or because of buildup during normal (though incorrect) operation of a woodstove. You'll be able to tell if your chimney is blocked, because your house will fill with smoke if you try to use the stove. Removing the blockage can be a great deal more difficult (and costly) than diagnosing it, sometimes requiring that the chimney be opened when more conventional methods fail.

When to Clean

Let's agree, then, that your chimney should be checked and swept, as needed, on a regular basis. How often? Every chimney sweep gets asked this question frequently, and one correct answer— whenever it needs it—rarely satisfies the person who poses it.

It's impossible to answer this question definitively because not only does each heating system differ from every other, the same heating system will differ from itself from one year to the next: Different wood and different weather conditions can lead to rapid creosote buildup in a normally clean-burning system. I've had some customers whose chimneys have needed monthly cleaning one year and just two cleanings the next.

So how often will your chimney need cleaning? According to the National Chimney Sweep Guild, whenever there are creosote deposits ¼-inch thick—less if it is glazed, since the glaze stores a great deal of energy. The only way to know when it is time to clean your chimney is to check it. For some people, whether because their heating system is difficult to examine or because they wisely refrain from climbing roofs, this will necessitate calling a good chimney sweep, and it should be done before the heating season and again when the sweep advises. Because most chimney sweeps are understandably reluctant to check the same chimney every few weeks, and because most woodburners are understandably reluctant to pay them to do so, I believe that it makes good sense for people to learn to check their own, if it's feasible. This is not in lieu of professional inspection, necessarily, but as a first line of defense.

Who can do this? If your heating system is simple—a straight flue venting either an open fireplace or woodstove or furnace—and you have no problem getting to the top of the chimney, then you're in business. Offset flues and difficult systems, such as stoves or inserts installed in fireplaces, are not impossible to check, but because you can't see the whole system, you can't inspect with certain results: If you see ¼-inch of creosote, you'll know that the chimney needs cleaning, but the absence of creosote where you can see doesn't mean that there's none elsewhere.

How to Check a Fireplace Flue

Remember, the flue must be straight, or nearly so, and, obviously I hope, the fire must be out. Dress in old clothes and gloves, and spread a drop cloth in front of the fireplace. Equip yourself with a strong flashlight or drop light, and carefully open the damper as far as you can—carefully because forcing some dampers may dislodge them. Before you look up, you should put on eye protection, because you will probably have to stick your head a good way into the fireplace to see up through the flue, and will thus be vulnerable to falling soot, creosote and mortar. You will also be exposing your lungs to dust and ash, so wearing a respirator is a good precaution to take. Shine your light into the wide area just above the throat, or damper opening. This is called the smoke chamber (FIGURE 2). It is usually wider than the actual flue and will not be tile lined. It may be, and should be parged—coated with a smooth layer of refractory cement—but more often the smoke chamber will consist of courses of brick tapering up to the flue opening. In an open fireplace, the smoke chamber is often dirty long before the flue.

Now direct your light into the flue. If the damper opens wide enough and the chimney isn't too tall, you may be able to see all the way to the top, in which case you can leave the ladder in the garage, if you wish. If you don't have a clear view of the entire flue, then it's up on the rooftop with you. Bring your light, of course, and look down the chimney. On a very bright day, you may have to put your face virtually in the flue to see anything. What you are looking for is creosote, which will be black or dark brown and, in a fireplace chimney, probably dusty or flaky. If you see ¼-inch of it anywhere, it's time for a cleaning. If you are uncertain what you're seeing, call a chimney sweep, and when he or she arrives, ask for some tips on what to look for along with the cleaning. While you're on the roof, by the way, take a moment to check the condition of the chimney's exterior, including its crown (the mortar that fills the space between the liner and the outer wall of the chimney) and the flashing: Cracks and separation in these places can lead to moisture damage.

How often you need to practice this dirty ritual depends on how much you use the fireplace: Their chimneys generally don't need cleaning as often as do woodstove chimneys, so if your use of the fireplace is occasional, you will probably be safe checking it once a year. Heavier use dictates more frequent checking, as does a fireplace with glass doors, as the doors reduce the amount of combustion air available to the fire and may result in slower, smokier burning. Do get a sweep to make the first inspection and, if necessary, clean the system, so that you're starting with a clean chimney and not leaving major problems undetected.

How to Check a Woodstove Chimney

Forgive me for saying so, but the fire in the stove should be out or you probably won't be able to see anything but smoke in the flue. You'll need the old clothes and gloves and flashlight. You'll also likely need a small mirror and a drop cloth, and you may need a metal pail and ash shovel, a screwdriver and some kind of respiratory protection.

Start at the base of the chimney—in the cellar, if you have one. What you're looking for is the clean-out, which should be seen as a small steel or cast iron door near the bottom of the chimney. If there's an opening but no door, you need a door, since the chimney will draw air through the opening, ruining the draft and adding an uncontrollable source of oxygen to any chimney fire. If there is no clean-out,

FIGURE 2.
Cross section of conventional fireplace

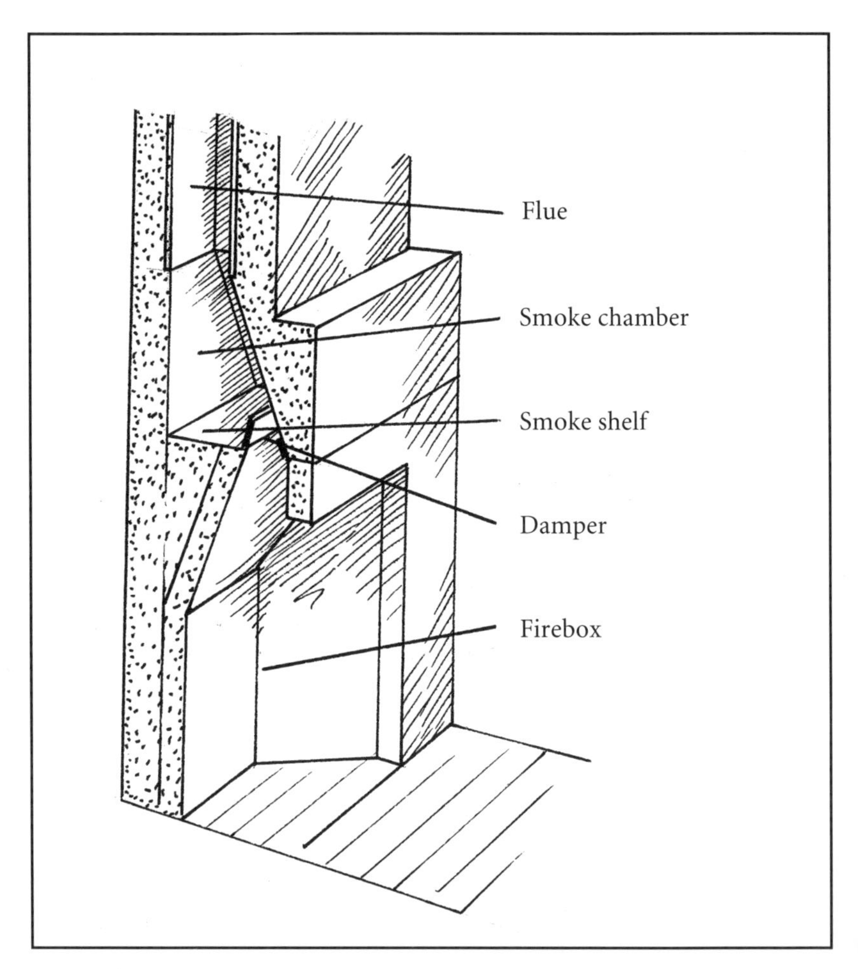

you need one (unless the flue doesn't extend to the base of the chimney, and the thimble provides reasonable access to the bottom of the flue), since creosote that can't be reached via a clean-out can ignite and, possibly, start a chimney fire. In either case, call a mason or a sweep who does masonry repair. In the meantime, you can still check the flue by removing the smoke pipe from the stove and checking through the thimble. A messy job.

If the chimney has more than one flue and more than one clean-out, ascertain which clean-out serves the woodstove flue. Note that if one of the flues serves a fireplace, the woodstove flue is probably offset where it passes the fireplace's smoke chamber, making it unlikely that you'll see anything from the clean-out.

OK—the flue is straight and has a clean-out. Put your respiratory protection on, spread your drop cloth, and open the clean-out door. You may have to pry gently with the screwdriver. If the clean-out is full of creosote, don't jump to conclusions; the chimney may be clean, since creosote will often fall off the inside of the flue by itself. Shovel the stuff into your pail—slowly, to keep the dust down, and when the clean-out is empty, remove a glove and put your bare hand in (the clean-out, not the pail). You should feel a draft—moving air. If you don't, reach into the clean-out above the door with the shovel or screwdriver and see if more creosote is hanging up, out of sight. If so, dislodge and remove it. Now hold the mirror in the clean-out and tip it toward you until you can see all the way up to the sky. (A bright day with the sun overhead is the best time to do this.)

If you can't see the sky, keep adjusting the mirror. If you still can't, either the flue isn't straight or it's blocked at some point, and you should call a chimney sweep.

If you can see the sides of the flue in your mirror, you're looking for the horizontal joints between the tiles all the way up, and clean, smooth tile surfaces. I wince when people say, "I checked it and it doesn't need cleaning; there's light coming through." You can have light coming through a chimney that contains enough creosote to burn a good-sized cruise ship to the waterline. You're not really home free if you see smooth tile, either, because it could be coated with glazed creosote, but if the flue is short enough and the day bright enough for you to ascertain that the tiles are pink or orange all the way to the top and not brown or black, you're headed in the right direction and can probably forgo the trip up the roof, unless the view from up there on such a fine, bright day is enough to lure you up to check the condition of the masonry, as discussed earlier. If you aren't certain that the chimney is either clean or dirty, head for the rooftop with your flashlight. Sometimes putting a drop light in the clean-out gives a very good view of the chimney's interior and, always, if you have enough extension cord, lowering a drop light slowly through the flue will give you a good idea of what's there. As with the fireplace flue, if you have any doubts concerning what you're seeing, call a chimney sweep. And do get a sweep to check and clean your chimney the first time. You can check it yourself when the sweep is done and thus establish a benchmark for your future inspections.

The last thing to check is your stovepipe. Some people try to get away with tapping the pipe with a fingernail—a clear, hollow sound being indicative of a clean pipe—but certainty can only be attained by taking the pipe apart and looking inside with a light. Check the thimble, too; creosote often builds up rapidly at this point in the system, particularly if the thimble is a foot or more long. This is also a good time to assess the condition of your stovepipe: Is it free of rust and fairly rigid? Is each joint secured by at least three sheet-metal screws so that the pipe can't shake apart in a pipe fire (same as a chimney fire, but it takes place in the pipe)? Again, it's best to go through the procedure the first time with a qualified chimney sweep.

Sweeping Your Own

I'm sure that many chimney sweeps will disagree, but I believe some people are well advised to sweep their own chimneys, particularly if they can secure the cooperation of a competent sweep who is willing to check their work from time to time. Who are these people? For the most part, they'll be among the people who can check their own chimneys, excluding those with fireplaces: Cleaning a fireplace chimney is obviously not impossible—I can do it, after all—but it does require equipment beyond brushes and rods, and it can be a filthy undertaking.

Why would anybody choose to clean his or her own chimney? With prices for a professional cleaning averaging, at this writing, $65-$70 per flue, many people of modest means who need several sweepings a year might be tempted to forgo some or all of them; better that they should hire the sweep for one and learn to do the rest themselves.

Alternatives

Before I explain how to sweep your chimney, I'll briefly discuss three other apparently inexpensive sources of chimney cleaning.

The first is a clean-it-yourself rig that doesn't require you to climb the roof. The ones I've seen consist of a frame and pulley at the top of the chimney, a brush in the clean-out, and a loop of cable which is attached to the brush and runs through the pulley at the top. By pulling the cable, you can work the brush up and down through the flue in the comfort of your basement. These can work, but they have several major problems:

1.) The cable in the flue interferes with the passage of smoke and contributes to creosote buildup.

2.) The acidic nature of creosote, exacerbated by the condensation in the flue, breaks down the apparatus fairly rapidly.

3.) Cleaning a chimney through an open clean-out will expose you to large amounts of carcinogenic dust and gases, and your house to a mess you'll not soon forget. It can be done, but it shouldn't be without a good respirator and a good vacuum cleaner.

The second chimney cleaning source I have in mind isn't as common now as it once was, but may still be available in some areas: Volunteer fire departments will sometimes sweep chimneys in exchange for a donation. This seems innocuous enough, and sometimes is, but it can also lead to real problems. Firefighters may know a great deal about extinguishing chimney fires, but there's no reason to suppose that they know anything about chimneys or chimney cleaning.

Two cases illustrate my point: I was called to clean a chimney that was malfunctioning; whenever the homeowner lit the stove, smoke would pour into the house. When I arrived, I found a stove installed in a fireplace, with the pipe passing through the damper and a metal plate closing the remaining space in the damper opening. The customer explained that she was puzzled by her suddenly smoky stove, since the local fire department had cleaned the chimney just a week before and had, in fact, cleaned it the preceding four years. The flue was not really clean, but neither was it dirty enough to restrict the draft. When I pulled the stove out of the fireplace, I found the source of the problem: The firefighters had evidently pushed the creosote down the flue year after year but had never questioned where it was going, which was, of course, into the smoke chamber and stovepipe, both of which were packed full. By the time I was done, I had filled three 30-gallon garbage bags with creosote.

The second case involved a department that offered a safety inspection of any local resident's wood heating system. A very well-meant offer, but potentially a very expensive one, since the firefighters did not know the acceptable clearances from combustibles for stoves and stovepipes. One installation I saw, which had been approved as safe by the firefighters, featured a woodstove installed in a fireplace a scant foot below a wooden mantel. The distance should have been three feet. I have heard of other departments cleaning chimneys with chains or feed bags full of bricks, either of which can break flue tiles. All in all, I think you'd be well advised to let the firefighters stick to firefighting, give them the donation they deserve for doing that dangerous and vital job, and have somebody who knows what he or she is doing sweep your chimney. The problem that the firefighters face is that you need to know a lot about chimneys and wood heating and have considerable experience with both to recognize problems in heating systems with which you are unfamiliar. You may be able to learn enough about your own system to do it justice, but going beyond that involves the sort of commitment to learning the trade that a chimney sweep should make.

The third cheap source of chimney cleaning is the various chemical cleaners on the market. These are usually either liquid or powder that you add to the fire. The problem is that they don't clean the chimney. Some may help convert glazed creosote to a brushable form of the menacing stuff and, in the process, cause some of it to fall off of the sides of the flue, but chemical cleaners are not a substitute for manual cleaning under any circumstances.

Now you're ready to learn to sweep your own chimney. If you thought that inspecting it was a nasty business, you're in for a real treat now.

I'm assuming that your system consists of a woodstove connected to the chimney through a thimble, rather than through a fireplace. I'll add that, even if your stove does connect via a thimble, if it shares a flue with a fireplace (the thimble being either over the mantel or on the side of the chimney opposite the fireplace), you should leave the job to a pro, because the smoke shelf will function as the clean-out in this case, and the removal of the creosote can be tricky and will possibly require a special vacuum cleaner.

I'll also acknowledge that a simple woodstove system can often be cleaned by a homeowner even if it has an offset flue; you may not be able to inspect it, but with the right equipment, you can sweep it.

Equipment

1.) *A chimney brush* sized to your flue. You'll need to measure the inside dimensions of your flue. Lined chimneys are standard nominal sizes (6-, 7- and 8-inch round, 8- by 8-inch and 12- by 12-inch square, and 8- by 12-inch and 12- by 16-inch rectangular tiles are very common) and brushes are made to fit them. Chimney brushes come in round wire and flat wire as well as polypropylene. For most people, the round wire brush is a good bet; it's less expensive than the flat wire models and will do an adequate job of removing powdery and dry, flaky deposits. If your system tends to generate hard, scaly creosote, you'll need to pony up the money for a flat wire brush; its stiff, chisel-like tines are effective in chimneys a round wire brush can't touch. If a flat wire brush is unable to remove the gunk in your chimney, it's past time that you called a sweep. The nylon and polypropylene brushes are the least expensive, but will only work on very dry, dusty deposits. Fireplaces and some insulated metal chimneys are often well-cleaned by these soft brushes (FIGURE 3).

2.) *Rods or rope:* Most brushes have a ring at one end and a threaded stem at the other (FIGURE 4). Chimney sweeps almost uniformly use flexible rods which attach to the stem of the brush and then to each other in series until the brush has traveled the length of the flue. Adaptors are available that allow you to attach a ring to the stem and use the brush with a rope instead of the expensive rods, either by hooking a rope to one ring and a weight to the other and lowering the

FIGURE 3. ***Common chimney brushes***

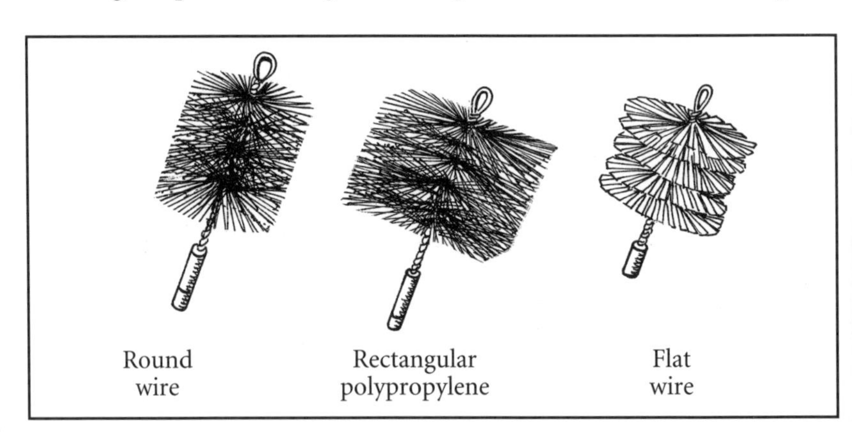

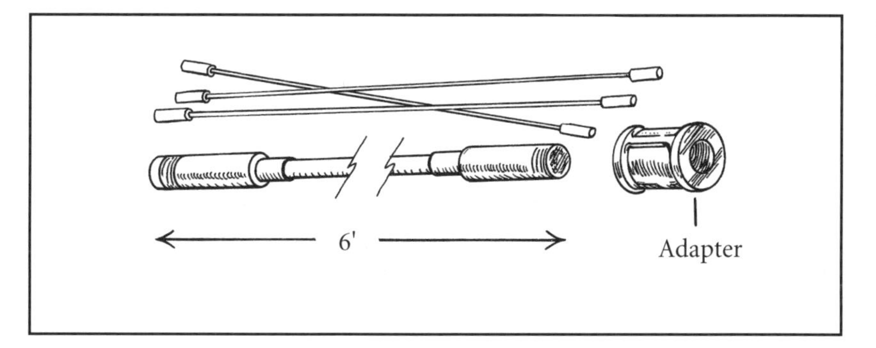

FIGURE 4. *Chimney rods and fittings*

brush down the flue, or by tying ropes to both rings and having a helper at the clean-out pull while the brush is lowered from above. The rope method can work, but not on a very dirty chimney or, often, one with a pronounced offset, so I tend to think that you are better advised to invest in a set of rods: They'll last you forever, unless you become a professional chimney sweep, and will amortize themselves quickly—you should be able to get 30 feet (a typical flue length for a two-story house with a basement) for under $100.

The most commonly used chimney rods are made of fiberglass. They come in three diameters—.480-inch, .440-inch, and .340-inch—and in lengths of 3-, 4-, 5- and 6 feet. Longer rods are cheaper by the foot, and shorter rods are usually handier for cleaning from the bottom of the chimney. Most sweeps rely on the heavy-duty .480 rods because they have enough rigidity to push a brush through a substantial creosote buildup. I think, however, that most homeowners will be well served by the .440 rods: They cost less, and if you find that they are inadequate for your chimney, you are either waiting much too long between cleanings or should be hiring a pro, or both. The .340 rods often work, but they're quite noodly and have difficulty moving a brush through more than a very light buildup. Their usual application is cleaning from the bottom of the chimney, which we'll discuss presently.

3.) *A shop-type vacuum cleaner* is optional; if, for example, your stove and clean-out are in a less-than-pristine part of the basement, you may find that a broom and dustpan are an adequate alternative. If you have to deal with creosote in an inhabited part of your house, however, you'll want a vacuum and are ill-advised to use your household model, since the very fine and corrosive dust that often comes out of a chimney may harm the motor.

4.) *A pail—preferably with a secure lid—and an ash shovel* for removing creosote from the clean-out. You don't know what trouble is until you've had a paper or plastic bag full of creosote rip halfway across the new white shag carpet in the living room.

5.) *A short-handled wire brush* for cleaning the stovepipe. If you have more than 2 or 3 feet of pipe, you might consider buying a round chimney brush to fit it: Attached to one length of rod, it will do a fine job of cleaning the pipe and will make it unnecessary for you to reach way in with a hand brush and fill your gloves with soot.

6.) *Whatever screwdrivers are necessary* to disassemble and reassemble your pipe. The most difficult and frustrating part of most chimney jobs is getting the screw holes in the pipe to line up when you're putting it back together. To minimize damage to my sanity, I disassemble the pipe only enough to remove it and clean it, and make a light mark overlapping any joints I do separate so that I can more easily match them up. I've also found (because my wife told me) that an awl makes short, easy work of lining up screw holes that don't quite want to go together.

A note on cleaning pipes: Because taking them apart can be such a hassle, some people bang on them, instead, using their hands or a stick. It is true that this forceful approach will often yield the satisfying sound of creosote falling through the pipe, but there are several problems with the technique. Where does the creosote go if you don't remove the pipe from the stove? If there are no elbows in the pipe and no baffles in the stove, it will probably fall into the firebox, which is okay, but otherwise, you haven't removed the creosote, you've just consolidated it at its angle of repose. Again, hitting the pipe doesn't usually loosen all or nearly all of the creosote; even if it falls into the firebox, you've probably left enough in the pipe for a good fire. Consider also that if you don't remove the pipe, you don't inspect the thimble, which is probably dirty. Finally, banging on the pipe may make it fall apart.

7.) *Personal protective equipment:* Creosote is a carcinogen, and skin contact and breathing it are to be avoided as much as possible. Obviously, cleaning your own chimney a few times a year will give you very limited exposure, and you will probably be reluctant to spend $700-$800 for a powered, positive pressure respirator, but you should consider a nonpowered respirator with cartridges rated for both dust and vapor. It isn't just the dust that is dangerous; the vapors are carcinogenic, as well, and this is why paper dust masks are inadequate.

Clothes that fit snugly at the sleeves and neck (turtlenecks—black turtlenecks—are perfect), and nonabsorbent work gloves and a hat should provide a reasonable amount of skin protection, particularly if you make it a point to shower and change clothes as soon as you've finished sweeping.

8.) *Drop cloths* are also optional, but spread around your indoor work area, they'll save you a lot of cleanup time.

9.) *Ladders.* It is not my intention to give a complete course on ladder use and safety—I'd have trouble passing such a course, anyway. I'm going to assume that, as I mentioned earlier, getting on your roof is no problem for you. If you've never done it before, I don't want to goad you into trying, particularly if your roof is slate with a 12/12 pitch atop a three-story Victorian. In lieu of the whole curriculum, I'll give you my favorite tips about working on roofs, minus a few that it would be irresponsible of me to pass on.

Working on the Roof

Unless your roof is flat or very modestly pitched, get a ridge hook and attach it to a single section of ladder long enough to reach from the eaves to the peak. Walking a ladder beats walking a roof any day.

Some roofs will not accommodate a peak, or ridge ladder either because the distance from eave to peak is too great or because they don't have peaks (e.g., a hipped roof). This may be enough to scrub the do-it-yourself project (unless your chimney can be cleaned from below), but if you think that you can walk the roof, I will, though doubtfully, rate the various common roofing materials for their walkability: 1.) Asphalt and fiberglass shingles usually offer good traction, even when wet. They can be icy and not look it, though, so proceed with caution in cold weather. 2.) Metal roofs are unpredictable: Old galvanized roofs are often walkable, and so are raised-seam roofs if you can help yourself up by grabbing the seams and avoid climbing in damp weather. I've run into a few metal roofs I could only walk by using the heads of the roofing nails as toe-holds, a practice I don't recommend and a tip I shouldn't pass on. 3.) Slate is usually fairly slippery and is fragile, besides. I put a peak ladder on slate roofs whenever it's possible, and walk carefully, near a gable end, holding the edge with one hand, when it isn't. Stay off of slate roofs when they are wet. 4.) Wooden shake roofs are much like slate, but are even more slippery when wet. They are also difficult to slide peak ladders on, because the thick shakes catch the top of the ladder. If you are walking a dry shake roof, watch out for the moss and lichen growing on it. They hold moisture and will be slippery when the rest of the world is drought-stricken. Why do people have shake roofs?

Don't lean a ladder on a freestanding chimney, even a new one. Mortared joints are intended to hold the chimney together, not to provide great lateral strength. Metal chimneys won't even provide the illusion that they can support you. When a chimney extends too far above the roof to be reached, I stand an aluminum step ladder against the chimney—not leaning, but vertically—and tie it to the chimney. I don't even do this when confronted with a tall metal chimney, however; I either take the chimney apart or clean it from below.

Have somebody foot the ladder when you climb. Most sweeps work alone and devise ways of using ladders without assistance—driving stakes and tying the bottom rung to them, for instance—but what working alone often means is taking chances.

When walking a roof, keep your weight centered over both feet by taking short steps, and keep a low center of gravity by staying crouched—the steeper the roof, the deeper the crouch. Holding the edge of the roof or even putting your free hand flat on the surface will help keep you from slipping.

You aren't climbing the roof for the fun of it, so you'll have to carry rods and brushes. It's always nice and often essential to have a free hand, so devising a way to carry your tools one-handed is necessary. Attaching the brush to one of the rods and then tying all the rods in a tight bundle would work.

The best shoes for roof walking are said to have gum-rubber soles, but I find that old running shoes work well, too; it's a good way to recycle them after their pronation control is shot. I'll hold off on identifying my favorite brand in anticipation of an endorsement contract.

Where do you keep the extra chimney rods when you're cleaning the chimney? I usually lay them across the top, between me and the flue. They aren't very likely to roll off, because they'd have to either go through me or the rod currently sticking out of the flue. It is a good idea, nonetheless, to keep people, cars, furniture and any other breakable objects a safe distance from the eaves in case something falls.

People assume that chimney sweeps use safety ropes, and some do. The problem is that there isn't much on a roof to which you can tie yourself; the chimney is probably it, and some chimneys would not support you if you fell, but would add to your considerable troubles by falling on you. If you are going to use a rope, tie it to the base of the chimney rather than the top or middle; the mortar is less likely to fail if your weight is suddenly suspended from the chimney. You might tie off on an obliging tree if one is handy. Just don't fall *toward* it if you must fall.

It's easier to climb up a roof than down, for some reason probably more psychological than physical. If you suspect that the ascent will tax your nerve to the breaking point, reconsider the project: You could get awfully cold and bored sitting on your roof waiting to be rescued by the fire department.

Cleaning from the Roof

I'm going to describe the steps in a routine top-down cleaning. This assumes that the chimney is not dirty enough to pose technical problems: no glaze or heavy, compacted buildup that would require special equipment to remove.

1.) Before setting up your ladders, secure your inside working area. This means that you should identify all openings in the flue you're going to sweep. There should only be two—the thimble and the clean-out—but, particularly in older houses, there may be more, and they may be hidden. I could tell you a nightmarish story about an open thimble behind a painting in a restaurant, and I've heard of open thimbles behind wallpaper that revealed their presence catastrophically when the chimney was swept. If you aren't positive that you know all the points where your chimney is breached, examine it as thoroughly as possible. Incidentally, any openings that are unused should be sealed with masonry to the thickness of the chimney and tile walls: In the event of a chimney fire, such openings can allow the fire to spread to the house and also provide an uncontrolled source of oxygen for the fire. Those tin pie plates commonly used to plug old thimbles have about as much chance of standing up to a 2,000-degree chimney fire as would a wad of paper towels.

Now that you've identified the openings in the chimney, make certain that they are sealed. A thimble with a tightly fitting stovepipe in it is often secure as is, but to be on the safe side, tape the joint where the pipe meets the thimble with duct tape. If your chimney is metal, tape the joint where the stovepipe meets the chimney.

Clean-out doors often aren't very tight and don't latch securely; tape between the door and its frame. Unused thimbles can be tricky to mask, particularly if they don't protrude from the face of the chimney. If tape will adhere to the chimney surface, you can put a plastic bag or crumpled-up newspaper in the thimble and tape a sheet of plastic over the opening. If you can't tape over the opening, pack the thimble well with paper or plastic, being careful not to push it into the flue, and brush very slowly to avoid creating a sudden vacuum which could dislodge your barrier.

One last precaution should be taken if your chimney has more than one flue: Not all the creosote falls to the bottom when a chimney is swept; much of the very fine stuff comes out the top. There is the possibility, if you have flues adjacent to the one you are sweeping, that some of the dust will go down and get into your living space through openings in that flue. Fireplace flues, probably because of their large diameters, are particularly susceptible to this problem, so before you start brushing, close the damper and, for extra security, consider covering the top of the flue with a board or plastic bag while you are working on its neighbor.

2.) Now to the roof. I like to make one trip to the chimney without my tools, mostly to establish my route and identify obstacles (slippery flashing at the eaves, inconveniently located TV antennas and lightning rods, wasp nests, etc.), but also to check the condition of the chimney. When you've checked the roof and chimney, get your tools ready to carry. Don't forget a flashlight to help you check your work, and your respirator. (Despite being outdoors, the top of a chimney while cleaning is in progress is an atmosphere choked with toxic gases and vapors.) Up you go.

3.) To clean, insert the brush, with a rod attached, in the flue and push it slowly to the bottom, adding rods as needed. You may hear creosote falling, and dust and soot will likely begin to come out the top of the chimney. If you encounter serious resistance, don't try to force the brush by it, since, if it is creosote, you may compact it and form a blockage. Instead, pull the brush back up the flue a few feet and gently try again to brush past the sticking point. (Incidentally, it isn't always easy to reverse a brush in a flue; I've had to rig a jack to accomplish this a few times by laying a stout stick across the top of the chimney, tying the rod tightly to it, and then using the stick as a lever.) If this fails, pull the brush out, give the dust a few minutes to settle, and use your flashlight to determine, if possible, what the obstruction is. If it is creosote, you can try, very carefully, to dislodge it with the tip of a bare chimney rod. Failing at that, or if you fail to identify the nature of the problem, you'll have to call a chimney sweep.

If you encounter no resistance, push the brush to the bottom of the flue, and then pull it back out and repeat the process until the chimney is clean, which you can probably determine with your flashlight. Note that clean chimney tiles may still be black, but they should be free of creosote buildups in excess of ¼-inch thick.

4.) Back in the house, spread your drop cloths around the stove, set up your vacuum cleaner, bucket and hand tools, disassemble the stovepipe and remove it from the stove. Now comes the judgment call: It's usually preferable to clean the pipe outside, but perhaps not if doing so entails carrying it across that white shag carpet. You can try wrapping the pipe in a drop cloth for the trip, but be careful. If you elect to clean it inside, turn your vacuum on and position the hose at the top of the bucket where it will catch airborne dust. Brush the pipe into the bucket with your hand brush slowly—speed creates turbulence and flying soot— until the deposits are gone. Now clean the thimble with your hand brush, vacuuming to control dust, and clean and vacuum the stove breeching (i.e., where the pipe enters the stove), baffles and smoke passages, if any. Be careful about vacuuming any part of the stove if it has been running recently; embers can last an astonishingly long time—three days or more under the right circumstances. Reassemble and reinstall the stovepipe and, after carefully folding your drop cloths and shaking them clean outside, head for the clean-out. By the way, it is often a bad idea to open windows or doors while the inside work is in progress; the fresh air may seem helpful, but the drafts it may create could disperse fine creosote through the house.

5.) To the clean-out bring your drop cloth, vacuum cleaner, pail and shovel, respirator and inspection mirror. Set the vacuum up, start it, and hold the hose slightly above the clean-out door, so the vacuum will help control the resulting dust. Carefully shovel the creosote into the pail and vacuum what you can't reasonably get with the shovel. It may be tempting to leave the creosote in the clean-out if you're in a hurry, but don't do it; a spark could ignite it, and damage to the clean-out door and chimney could result.

Now inspect your work with a mirror: You should see a clear flue all the way to the top, unless the chimney has an offset, in which case you should feel for a draft. If you have neither draft nor view, you've possibly left a blockage of creosote in the flue, and you'll need to return to the roof unless you can reach it from the clean-out. This is a frustrating experience and one you shouldn't have unless you clean your chimney too infrequently. If I have any doubt that I've brushed clear to the clean-out, I push the brush as far down as I can, then I get off the roof before finishing the cleaning and check the clean-out to make certain that the brush is there.

6.) Your last step is to double-check: Is the clean-out door tightly closed? Is the stovepipe properly reinstalled with three screws at each joint? Are any repairs needed? OK. You're done. It wasn't so bad after all, was it?

Cleaning from the Inside

Many chimneys can be cleaned from the inside or, more accurately, from the bottom. Fiberglass and polypropylene rods are flexible enough to push a brush through most fireplace damper openings and many clean-out doors and stovepipe thimbles, and since this method will keep you off the roof, why not use it if you can?

The answer is that without a special vacuum cleaner, the dust generated by from-the-bottom chimney cleaning is very difficult to control. This won't matter too much if your clean-out door is outside, and it may not matter if you can effectively mask the work area, but unless your roof is nearly impossible for you to climb, cleaning from the top will probably be easier. Also, it is important for you to periodically inspect the condition of the chimney from the roof line up, and it is often necessary to clean chimney caps, particularly those on metal chimneys, since they seem to have a penchant for collecting creosote. Having said this, I'll acknowledge that your situation may demand that routine cleanings be done from the bottom.

How can you tell if it's possible in your case? By looking, first, then trying. A clean-out which is not elevated above floor level will usually be impossible to start a brush through, even if you use the light .340-inch rods, because it would demand more flexibility of the rods than they could have and still push a brush effectively. The same is true of a flue that is recessed more than a few inches from the clean-out or thimble opening. If the only contra-indication to cleaning from the bottom is the elevation of the clean-out, you may be able to clean through the thimble, if the brush will fit. In any case, you won't know that it is possible to clean from the bottom until you try to start the brush in the flue. This is seldom easy to do; you'll probably have to reach in the clean-out or thimble and push the bottom of the brush up as you apply pressure with the rod.

If you succeed in starting the brush up the flue, here's the procedure for finishing the job:

1.) Secure all chimney openings save the one you're going to use. You'll be much happier if you can mask that one, too, since a lot of dust will probably billow out of it otherwise. One way of doing this is to start the brush and rods, then tape a piece of plastic or heavy paper with a slit cut for the rods over the opening.

2.) Move cars, people, lawn furniture, etc. away from the eaves in the proximity of the chimney: Large amounts of creosote may be pushed out, and sometimes loose bricks from the chimney top can be dislodged: Both are subject to the laws of gravity.

3.) Set up your vacuum and drop cloth at the clean-out (if that's where you are working) and position the vacuum hose near the opening.

4.) Start the vacuum and push the brush slowly up the flue, adding rods as needed. Retract and repeat. Resistance should be dealt with as discussed before, but there is less concern about compacting creosote when cleaning from the bottom, so your main worry is that the resistance is a broken tile or an offset in the flue, in which case you could do damage by applying excessive force.

5.) When the chimney is clean, follow the procedure previously described for dealing with the stovepipe and clean-out.

Hiring a Pro

If you can't clean your chimney, don't want to clean your chimney or are having a problem with your chimney, it's time to hire a chimney sweep—past time, perhaps. And, just as you need to be selective in choosing a doctor or an auto mechanic, you should have more criteria for picking a sweep than price. To expand on this comparison a bit, you will derive the most benefit from a good doctor who knows your health history, or a good mechanic who knows your car, or a good sweep who knows you and your heating system. A sweep who sees your chimney year in and year out will be able to spot changes and diagnose performance problems more quickly and accurately than will somebody you hire on a one-shot basis.

Here are some guidelines for picking a chimney sweep—an important decision, after all, since it might well affect the safety of you, your family and your property.

1.) Ask other woodburners in your area whom they use. If you live in a small town or in a rural area, your choices may be limited, but most people in the woodburning belt will be able to find several—

and perhaps many—chimney sweeps from whom to pick. Getting some consumer reactions is a good first step.

2.) Look in the Yellow Pages. This will give you a pretty good idea of how many sweeps work in your area, and the ads themselves may reveal or imply some important information: Is the sweep certified by the National Chimney Sweep Guild (NCSG)? What services does he or she offer? Is the ad informative, or does it emphasize sizzle rather than steak—top hat and tails rather than service?

3.) Let's return to the question of certification. In order to be certified by the Chimney Safety Institute of America (CSIA), the educational arm of the NCSG and the only national organization that specifically certifies sweeps, a candidate must pass a rigorous written exam based upon a manual prepared by the CSIA and upon the safety codes developed by the National Fire Protection Association. Certification does not guarantee that a sweep will have the mechanical proficiency to clean your chimney, however, nor is it a guarantee of honesty. And the lack of certification is not proof positive that a sweep lacks essential technical knowledge. What certification does indicate is that the sweep has mastered a comprehensive body of technical information, and while I acknowledge that there are excellent sweeps who are not certified, I would advise you to hire one who is; the trade is not regulated, and so the NCSG's certification program provides the only standard besides reputation and personal contacts that a consumer can apply in making his or her choice.

4.) When you've got a short list of prospective sweeps, it's time to make some phone calls. You'll want an estimate of the cost, certainly, and you might also ask what the sweep's cleaning method is: If it's scrubbing the flue with a Christmas tree or a live goose, you've heard enough and can hang up with no fear that you are missing out on a bargain. Wire brushes and vacuum cleaners are the standard tools of the trade, and any competent sweep will have them.

You also need to ask what you'll get for the quoted price: Does it include cleaning the stovepipe? If you have a fireplace, does the sweep clean the smoke chamber and behind the damper? (Don't hire one who doesn't.) Some sweeps charge extra for answering questions—consultation—so you should establish whether or not that is the case before you hire.

5.) Be sure to ask whether the sweep has liability insurance, and don't hire anyone who doesn't. If you have any doubts, ask the sweep to furnish you with a certificate of insurance.

I've spent a lot of time on the subject of chimney cleaning because it is critically important to anyone who burns wood. I'll say again that wood heating is an active pursuit, not a passive one, and that even if you don't clean your own chimney, learning as much as you can about the process will serve you well.

TROUBLESHOOTING CHIMNEY PROBLEMS

For all its pleasurable aspects, heating with wood can lose a bit of its luster when problems develop; when, for instance, your stove won't burn hot and fills your house with smoke whenever you open the loading door. Any kind of heating system can malfunction, of course, but wood-fired systems, because of their variables, can pose problems that are not obvious as to cause or solution. Sometimes—often—dealing with such problems is best done with professional help, but there are many cases where a homeowner can diagnose and remove the cause of the troublesome stove.

The first thing to recognize is that, in a vast majority of cases, a performance problem in a woodburning stove or fireplace is a problem with the chimney. An elementary understanding of how chimneys work should help you figure out why they sometimes don't.

How a Chimney Works

A chimney that is functioning as it should is a column of warm air (or gas) which, as it rises, causes a condition of low pressure—i.e., suction—at the bottom of the chimney. This suction draws air through the appliance where it is heated, causing it to rise and so perpetuating the condition. The suction is known as draft. Actually, most chimneys will have draft even when their appliances aren't operating, because the air in the closed column will still tend to rise, and since the pressure at the bottom of the chimney will usually be lower than the pressure in the room outside the chimney. Nevertheless, the hotter the gas in the flue, the faster it rises, and the greater the resulting draft.

Besides having a column of rising gases, a successfully operating chimney will be able to handle a sufficient volume of gases. In other words, the chimney that vents your fireplace must be large enough to move all the smoke produced by the fire up and out of the house. Volume is not independent of draft: To a point, the greater the draft, the greater the volume of smoke the chimney can process.

Diagnosing Chimney Problems

Because of the variables in wood heating systems, it's important that you collect data when a problem develops so that whoever deals with it has a basis for narrowing the field of possible causes. Some questions you need to ask and answer include:

1.) Did the problem just manifest itself, or is it chronic?

2.) Is it constant (the stove always smokes)?

3.) Does weather seem to affect it? What weather?

4.) Does the stage in the burn cycle seem related to the problem?

Now we'll apply what we know to some sample chimney problems, both short- term and chronic.

Short-Term Problems

By a short-term problem, I mean one that recently developed in a usually trouble-free system. The most common of these is a stove which, more or less suddenly, smokes when in operation, particularly when the loading door is opened. Most often the cause is creosote buildup sufficient to reduce the chimney's volume capacity to the point where it can't handle the smoke produced by the stove. Frequently, the chimney will actually be blocked in one or more places. An attentive stove operator, particularly one who habitually checks his or her own chimney, would usually notice this problem in an early stage, but creosote buildup can be very rapid under the right circumstances. The immediate solution, of course, is to clean the chimney. The long-term solution involves finding out why the system is subject to such a heavy creosote buildup and trying to remedy the situation. These causes might include:

1.) Sizing. The stove is too large for the space it is heating, and consequently is usually operated without enough oxygen for an efficient burn.

2.) Weather. Relatively warm winter weather can help make chimneys dirty rapidly: People slow down their fires under these conditions, leading to incomplete combustion and a smaller than usual temperature difference between the gases in the flue and the outside air, which results in decreased draft. A solution is to build smaller fires but not overly restrict the oxygen supply. You might also open a window or two to cool off the house, or you might simply not use the stove during warm weather.

3.) Bad fuel. Wet or unseasoned wood tends to burn incompletely. The solution is to either find dry wood or take pains to burn the wet wood hotter.

In a relatively small number of cases, a stove may start smoking for reasons other than a creosote blockage. These might include:

1.) Weather. Again, warm weather—particularly if it's cloudy, calm and damp—leads to sluggish draft. I've encountered stoves that smoked but had clean chimneys and pipes and worked fine in colder weather. There are likely to be other factors involved in a case such as this, but operating a woodstove in 50-degree weather can lead to trouble.

2.) A mechanical problem—perhaps the catalytic converter is plugged (in which case the stove should work fine in its non-catalytic mode) or a broken chimney tile is obstructing the flue, or a chimney cap is plugged. I once responded to a call from a customer with a smoking stove. Puzzled at finding the chimney and pipe clear, and seeing no obvious reason for the problem, I finally removed the top of the stove and found that a sheet of paper—kindling from an earlier fire—had been drawn into the baffle where it blocked the smoke's path to the chimney.

A second short-term problem is a stove that has started to burn sluggishly. The chances are good that this is the same problem as the smoking stove but at an earlier or less severe stage. It could, however, signal a dead catalytic converter, a worn-out thermostat on a stove with a thermostatically controlled draft, or even an excessive buildup of ashes which restrict air intake.

A third problem is the opposite of the second: A stove which suddenly burns too hot and is difficult to control. A likely cause is overly dry wood which needs very little of the fire's energy to drive off moisture. The solution is to mix the super-dry wood with less dry wood, rather than shutting down the stove's air inlets, since dry wood in an oxygen-starved fire produces a lot of smoke and creosote.

A second possible cause for a runaway stove is worn-out stove door gaskets. This condition allows the stove to draw air on demand, even when the air inlets aren't wide open, and will result in very hot fires. A loading door that latches very easily and/or has any play in it when closed usually has worn-out gaskets. The solution is to replace the gaskets.

Chronic Problems

Long-term problems are often more difficult to diagnose and more expensive to solve than short-term problems because they frequently are indicative of a heating system with mismatched components or ill-conceived design.

A common problem is a stove or fireplace that smokes during windy weather. This may be a result of turbulence at the top of the chimney, or a stove may smoke because the chimney is on the side of the house hit by the prevailing wind, which results in higher pressure at the top of the chimney than in the house, reversing or negating the draft. In either case, increasing the height of the chimney should solve the problem—in the first case because it would bring the top of the chimney above the turbulence, which is commonly caused by nearby, taller objects (the roof, trees, other chimneys), and in the second case because the raised top will be above the wind-created area of high pressure. A chimney should be at least 2 feet taller than any object within 10 feet of it, and should project from the roof at least 3 feet. These are minimum specifications, though, and may not be sufficient in some cases.

If the smoking is caused by turbulence, a chimney cap may solve the problem inexpensively by deflecting wind blowing down on the top of the chimney. A fireplace that *always* smokes likely does so for one of three reasons:

1.) *The flue is too cold* and hasn't enough draft to be self-starting. This situation will usually be associated with an exterior chimney and is easy to diagnose. If the smoking stops once the fire is well established, and you can overcome the smoking by holding a flaming torch above the damper opening prior to starting the fire, a cold chimney is your culprit. The solution is to always use the torch when starting the fire. Don't shoot your architect.

2.) T*he flue is too small* for the fireplace opening. Again, this is fairly easy to diagnose: Get a board or a piece of cardboard as wide as the fireplace opening, and kindle a fire. When it begins to smoke, hold your board just below the lintel and gradually lower it: If the smoking stops, you may have found the cause. To confirm the finding, measure the flue (see note above on measuring flue area), and determine its area, likewise the fireplace opening. The flue's area should be at least one-tenth that of the fireplace. If it isn't, your solutions include lowering the lintel permanently or installing glass doors. Don't shoot the mason.

3.) The fireplace doesn't have enough air available to maintain the necessary flow. Tight, modern houses in particular often don't allow enough air infiltration to feed a hungry fireplace. To diagnose this cause, kindle a fire. When it begins to smoke, open a window. If the fireplace stops smoking, insufficient air is the problem. The solutions are to always open the window, install outside air returns or install glass doors. It should be noted, however, that ready-made outside air returns are often too small to supply enough air for a fireplace. The size of the *damper opening* is the size your outside air returns may need to be in order to effectively eliminate the smoking.

A stove that always or often smokes or is sluggish may behave this way for a number of reasons. We're assuming that the chimney and pipe are clean and that the stove is not defective. The things to check are:

1.) *The chimney flue.* If it's too large in diameter, too short or too cold, or a combination of the three (a situation common to new stoves installed in fireplace flues), the system may not provide a strong enough draft for the stove. The solution is to furnish the stove with an appropriate chimney, either by lining and insulating the existing one or by replacing it.

2.) *The stovepipe.* If it's too long or has too many elbows (more than two), it may allow the smoke to cool too much, resulting in a weak draft. If the pipe is smaller in diameter than the stove's collar, it will overly restrict the flow. The solution is to install properly sized pipe in a less restrictive configuration, which is often easier said than done.

3.) "*Stack effect.*" In some situations, particularly a tight house with a stove in the basement or on the ground floor, the house can become a more efficient chimney than the chimney itself, causing the smoke to flow down the chimney and into the house. This will most often happen when the stove is burning slowly and there are windows or doors open on the upper level of the house but not the lower.

What happens is that the warm air in the lower part of the house rises in compliance with the well-known physical law, its buoyancy abetted by the draft created by the open window. This causes low pressure to develop in the bottom of the house—lower than the pressure in the bottom of the chimney when the fire is burning slowly—and, presto, your house becomes a smoky chimney. To determine if this so-called "stack effect" is your problem, close any windows and doors on the upper level of the house and open a window on the lower level. If the smoking stops, stack effect is the culprit. You may be able to solve it by not opening upstairs windows and avoiding smoldering fires, but more likely will have to provide a source of outside combustion air for the stove.

Your stove may back-puff occasionally for reasons that are not wind related. You'll recognize a back-puff because it is sudden, sharp and often followed by several more. What's happening is that the wood burning in the stove is producing combustible gases, but the oxygen supply is insufficient for them to ignite or exit the flue. As the stove draws in more air, it provides enough for the stationary gases to suddenly ignite, which they do, with an explosion that pushes smoke out through the stove's air vents. The vacuum created by the explosion pulls in more oxygen, causing another explosion if there's enough gas available. The management of the stove is what causes back-puffing—specifically, too abruptly and completely restricting the oxygen supply to a large, hot fire. The combustion is hot enough to produce gas, but there isn't enough oxygen to burn it. The solution? Give the stove more air.

Chimney Fires

A chimney fire is not so much a performance problem as it is the result of one or more performance problems. Because a homeowner's initial reaction to a chimney fire may well determine how destructive it becomes, it is well worth quickly reviewing the correct procedure to follow.

Detection of a chimney fire is not usually a problem: It will probably announce itself with a prolonged roaring noise, smoke and odors in the house and thick, dark smoke and/or sparks and flames coming out the top of the chimney. Some chimney fires are not so dramatic, probably because they haven't enough fuel or oxygen to really take off, but all chimney fires are potentially destructive and should be taken seriously. To people who regard them as a harmless way to clean a chimney, I can only say that physicians used to bleed people who were ill, too; all of the available objective evidence indicates that both practices are foolhardy.

For people with chimney fire paranoia, there are detectors available that indicate when the flue temperature is conducive to a fire.

Should you have a chimney fire, the first two things to do, simultaneously if possible, are:

1.) Shut off the fire's air supply as completely as possible by closing all air intakes on a stove or covering the opening of a fireplace with something nonflammable.

2.) Call the fire department. Many people don't take this step, probably out of embarrassment or to spare the firefighters, but I've been assured by every firefighter I've ever asked that they want to be called. It's much easier to contain a chimney fire than it is to extinguish a structure fire. If you are alone when the fire occurs and can't take Steps 1 and 2 simultaneously, I believe that you should shut off the air supply before making the call, unless the fire has already spread to the house or cutting off the fire's air will take you too long. The reason is that depriving the fire of oxygen in the beginning may be the critical factor in keeping flue temperatures from skyrocketing and destroying the chimney. *Never* open the stove door or the clean-out after shutting off the air: To do so is to supply the fire with a sudden abundance of oxygen and will likely result in an explosion.

When you have shut off air to a chimney fire and called for help, the final step is simple but hard: Prepare to evacuate the building, should that become necessary.

Water Problems

Again, not a problem with performance, but a potentially damaging problem, nonetheless. Water, added to the residues of combustion found even in clean chimneys, can speed the deterioration of masonry and metal flues. Water penetrating a masonry chimney, either through the crown or the porous brick or block exterior, can result in significant damage, particularly when it freezes. Not all chimneys seem prone to either problem, but if you notice water in your system—it may show up in the clean-out or the fireplace—or see signs of water damage such as rusty metal components or flaky, crumbling masonry, it behooves you to take remedial action.

Water entering the flue is usually easily stopped by installing a cap (a removable one, please, to keep your sweep happy). Water that seems to be penetrating the chimney's exterior may be coming through the crown, the masonry itself, or the flashing. If the crown is cracked, have it repaired or replaced. If the flashing isn't securely sealed, repair it by caulking it with silicone, or replace it if necessary. If these steps don't cause the problem to abate, you'll need to consider waterproofing the chimney itself, a relatively inexpensive service offered by many chimney sweeps.

Water-related chimney problems can be vexing to diagnose and solve. It may make sense and save you money in the long run to get professional help if you suspect that your chimney has sprung a leak.

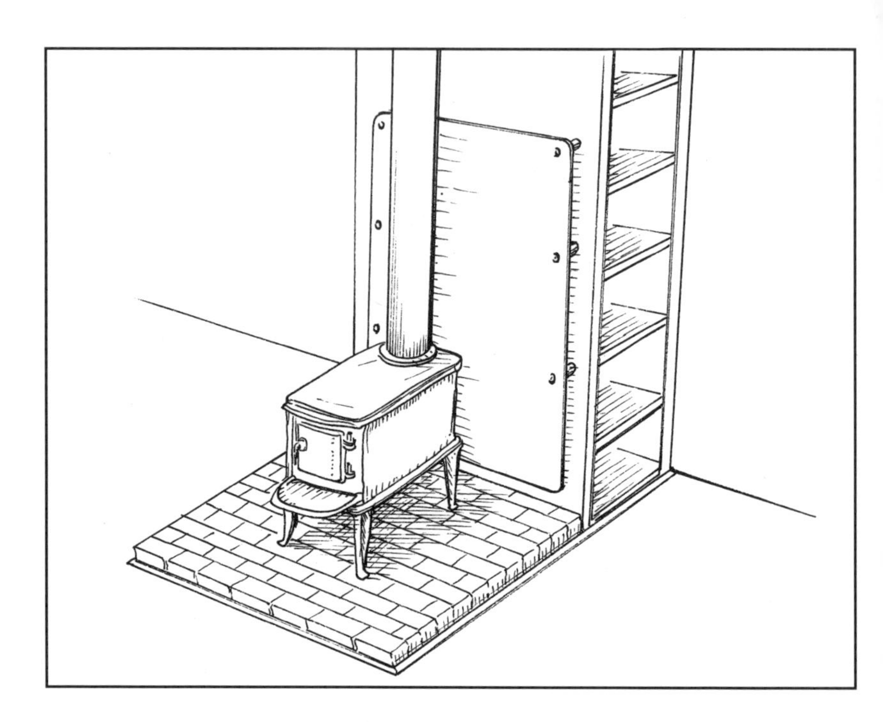

Chapter 7

Stove, Furnace or Fireplace?

It's time to ponder variables again: Wood heat is available in three basic forms—stoves, furnaces (in which group I include boilers) and fireplaces. In addition, each main group has subgroups, and there are a pair of interesting heaters—pellet stoves and masonry heaters—that don't neatly fit in any of the three categories. To determine which choice is best for you, consider again the components of your heating system, potential or existing: You, your house, your chimney, your location and the heating appliances under consideration.

Whether you choose a stove, fireplace, furnace or masonry heater, the rest of your system will affect and, to varying degrees, be affected by the decision. For example, as we saw in an earlier chapter, the design of your house (e.g., small, very well insulated, with lots of interior partitions) may make a woodstove impractical unless you undertake large-scale renovations. Your first step is to consider what you want from wood heat: esthetic enjoyment? An open fireplace is probably the purest embodiment of that. Lower fuel bills? The more of your heating requirement that is met by burning wood, the more money you will save. A wood furnace or stove can sharply reduce or eliminate your dependence on other heating fuels. Or do you want both the esthetic charm of woodburning and the savings? You might lean toward a wood boiler in your basement and a fireplace in your living room, or perhaps a handsome woodstove with glass doors would satisfy both needs. The point is to have a wood heating system that does what you want it to do and works the way it is supposed to—a system in which all parts are well matched. To that end, I'll discuss the different types of wood heaters with an eye to their strengths, weaknesses and best applications.

WOODSTOVES

Woodstoves have certainly changed: The 1990 advent of the EPA's final stage of emissions standards has brought us cleaner burning, more efficient and more expensive stoves. Their place in the wood-burning spectrum remains unchanged, however: A stove is a space heater capable, under some circumstances, of heating an entire house and providing, ideally, some esthetic satisfaction.

Among stoves, there are four subgroups: old airtight, older non-airtight, new catalytic and new high-tech noncatalytic. The old airtights—the "black boxes" of the 1970s—are still available secondhand, as people upgrade to the more efficient modern models. A good airtight can do a very effective job of heating and will hold a fire overnight, but will produce dangerous amounts of creosote and unconscionable quantities of air pollution if burned incorrectly.

Older, nonairtight stoves—most of them antiques at this point—are less readily available, but not impossible to find. Often ornate, these stoves, because their combustion air supply is hard to control, tend to produce less creosote than do airtights, but because so much of the heat that they produce goes up the chimney, they use more wood to heat the same space, as well, and cannot be counted upon to hold a fire overnight.

Catalytic stoves employ a catalytic converter or combustor to reduce air pollution, with the happy side-effect of producing more heat per cord of wood. The converter, a ceramic honeycomb coated with either platinum or palladium, reduces the ignition temperature of the tars, vapors and other organic compounds that make up woodsmoke. When the converter is preheated to light-off temperature (500-700 degrees Fahrenheit) and the gate that allows the smoke to bypass the catalyst during preheating is closed, the catalyst burns the smoke. The result is that creosote formation and air pollution are reduced by as much as 90%. In addition, because of the extra heat the burning smoke produces, wood use is frequently reduced by one-third. Catalytic stoves generally hold their fires as long as airtights, but they require the operator to pay attention to properly preheating (but not overheating) the catalyst, and they need new catalysts—at $100 and up a pop—after ten to twelve thousand hours (two or four seasons of regular burning).

The noncatalytic high-tech stoves rely upon secondary combustion instead of catalytic converters to achieve the clean burn mandated by the EPA. They attain secondary combustion (in some cases, three or four stages of combustion) by guiding the smoke to hot spots in the firebox and, usually, introducing a fresh supply of combustion air, with the result that temperatures in excess of 1,000 degrees burn the smoke. High-tech stoves do not burn quite as cleanly as catalytic stoves working at peak efficiency, but they may well do so over the course of several seasons, since the cat stoves' performance probably declines near the end of their converters' lifespan. High-tech stoves, like catalytic stoves, hold fires overnight and require that the operator pay attention to maintaining the correct temperature, particularly at the beginning of the burn cycle.

Best Applications

Stoves are not central heating equipment, so they should be located where the heat is most needed—in the living space. If your house was designed to be heated by a stove or has, by fortuitous circumstance, few interior partitions and is compact rather than sprawling, you likely can do all or almost all your heating with a stove. Most houses don't lend themselves to this, however; a stove large enough to heat a good-sized ranch-style house would probably make the room in which it was located unbearably hot. In most houses, then, the most realistic role for a woodstove is as supplemental heat.

Advantages

1.) *Price.* Although new stoves typically cost $1,000-$1,400 or more, they can amortize themselves rapidly. Corning, Inc., manufacturers of catalytic converters used in stoves, calculates that heating a 2,000-square-foot home with a catalytic stove will cost $325 less per season than it would with an oil furnace. The savings would not be as great with other types of stoves, but would still be substantial.

2.) *Looks.* A furnace represents woodburning's utilitarian extreme—so ugly that it's hidden in the basement, but it heats the whole house—and a fireplace represents the esthetic extreme: beautiful and mesmerizing, but, in many houses, so inefficient that it represents a net heat loss. Woodstoves reside between the extremes: You may not be able to heat your entire house with one, and it may not provide the ambience of an open hearth, but it fulfills both needs well enough. As I write this, in fact, I'm enjoying the company of my soapstone stove; it has a large window, so I can see the flames, and it is heating the entire house. It seems to me to be a very good compromise.

3.) *Adaptability.* Installing a woodstove in a house is rarely as easy as it might seem, but it usually is less expensive, disruptive and time-consuming than installing a furnace or fireplace.

Disadvantages

1.) *Mess.* Woodstoves are messy. Dust, sawdust and bits of bark constantly try to take up residence in the vicinity of a stove, leaving you three alternatives: live with the mess; devise a way to minimize the mess and deal with it easily (woodbox near the stove, hearth broom and shovel handy); or be driven progressively more crazy by it until you give up the stove. I've seen all three solutions practiced, and recommend the second, but I suspect that true neat freaks will find themselves dreading the heating season if they have a woodstove.

2.) *Work.* Stoves require active, knowledgeable participation by their owners. So do fireplaces and furnaces, of course, but perhaps not as much.

I don't consider this a weakness, by the way, at least not until the end of March, but I don't doubt that people who feel strongly that heat should be a convenience will disagree. It's not, I hasten to add, that you must constantly baby and coddle a woodstove, but neither can you simply set a thermostat and allow anonymous technology to do the rest.

FURNACES AND BOILERS

It is technically incorrect to lump furnaces, which heat by circulating warm air, with boilers, which circulate heated water. From the homeowner's standpoint, however, the appliances have much in common: They provide central heat, they are not located in the living area, and aspects of their operation, such as control of combustion air, are typically more automated than are a woodstove's. I have seen conventional basement-installed woodstoves that use the oil furnace's ductwork to circulate warm air through the house, and I suppose that the addition of a plenum to gather heated air transforms these units into primitive furnaces, but what I'm describing here is appliances designed specifically to be central heaters.

As to which is best, furnace or boiler, each has its advocates: A furnace may be cheaper to install, but a boiler heats water, a more efficient heat-transfer medium than air, and doesn't blow air into the living space; moving air feels cooler than still air of the same temperature. In the end, your choice in an existing house will likely be based upon the oil or gas-fired heating system in place: If it's forced hot air, you'll probably be loath to install expensive new plumbing. One note of caution: The installation of a wood-fired boiler or furnace, even if it's an add-on (one that uses the existing circulation system), should only be done by a thoroughly knowledgeable professional. Incorrectly installed furnaces and boilers may function poorly and can be subject to dangerous heat and/or pressure buildup. Also, the woodburning appliance should not share a chimney flue with the oil or gas-fired appliance unless it is designed and approved for that sort of installation.

Best Applications

Large or sprawling houses that cannot be effectively heated with a stove usually can be with a furnace or boiler. Also, furnaces and boilers don't require a lot of operation. You do have to stoke them, but modern units pretty much do the rest (except ash removal) based on the thermostat setting. That relative convenience makes furnaces and boilers appealing to those who seek the economic benefits of woodburning but who don't feel the need for a woodstove as the centerpiece of the living area.

Advantages

1.) *Location.* Furnaces and boilers are usually in the basement, so that's where the mess associated with woodburning will be confined (unless you have to carry wood through the house to get it to the basement—a situation that I expect you'd rectify quickly).

2.) *Effectiveness.* Properly installed, a wood furnace or boiler will heat the entire house fairly evenly, unlike a woodstove.

3.) *Convenience,* already discussed.

Disadvantages

1.) *Price.* Furnaces and boilers are generally much more expensive than stoves—several thousand dollars and up, not counting installation or any ductwork or plumbing that may be necessary. If your house can be heated with a stove, it's unlikely that the furnace will ever pay for itself by comparison. If your only alternative is a fossil-fuel-fired furnace, however, the wood-fired model will amortize itself in time.

2.) *Diffuse heat.* The disadvantage of central heating is the disadvantage of the wood furnace and boiler: There's no place to go to get warm, unless you care to set up a rocking chair in your basement. You wouldn't in mine.

3.) *Dependence on electricity.* Many furnaces and boilers need electricity to work, since they depend upon forced drafts, pumps or circulating fans. This could be a problem if you are subjected to frequent power failures.

4.) *Lack of charm.* I've never seen a beautiful furnace or boiler: Handsome in a utilitarian way, perhaps, but never soothing to gaze upon.

FIREPLACES

The National Chimney Sweep Guild classifies fireplaces as "entertainment-oriented appliances," and this is a generally fair assessment. We've mentioned that fireplaces often remove more heat from the living space than they provide. (Somebody once suggested, only half in jest, that for home heating, most people would be better off playing a tape of a crackling fire in their VCR than lighting a fire in their fireplace.) How, then, did our ancestors manage to heat their homes with fireplaces? There are several answers, the most important of which is that our ancestors' homes weren't very warm. A second answer is that their fireplaces probably did produce a net heat gain, because they weren't sucking furnace-heated air up the chimney or using it for combustion. Knowing this wouldn't have made you any warmer if you'd been trying to read by candlelight in the back bedroom of one of those colonial houses on a February night, but it does point to several improvements you can make in your conventional fireplace.

By installing glass doors and closing them when a fire is burning as well as when it's not, you sharply reduce the fireplace's demand for combustion air as well as its tendency to siphon warm air out of the house even when it isn't in use. (Most dampers are far from airtight, and none are insulated.) A second important step you should take if you have or are contemplating having a fireplace is to provide cold air returns. (But see the note in Chapter 6 on outside air returns.) This allows the fireplace to draw its combustion air from outside; the air there is, one would hope, more plentiful than it is in your house, and you haven't, one would also hope, paid to heat it.

An option you might consider, if you haven't already got a fireplace and can't bear to be without one any longer, is to have yours built to the 200-year-old specifications of Count Rumford. The Rumford fireplace, long-neglected but making a comeback, differs from a typical low-linteled, deep fireplace in that its high, wide and very shallow firebox and low fireback effectively radiate its heat into the room. The placement of the chimney flue and the design of the throat (damper opening) assure a draft strong enough that, even with the fire built where it should be—partially on the hearth—the Rumford won't fill the room with smoke.

While not nearly as efficient a heater as a stove or furnace, a Rumford supplied with outside air may provide a slight net heat gain while allowing you to see and hear the fire. You can have your cake and eat some of it, too.

It is important to note that a true Rumford has very exacting specifications, which must be followed if the fireplace is to live up to its reputation. Some "Rumford-style" fireplaces—look-alikes that alter the design proportions—are probably not any more efficient than conventional fireplaces and have the added disadvantage of being unsuited to the installation of glass doors, since the proximity of the fire in the shallow firebox will break the glass.

Best Applications

You aren't going to heat your whole house with one, so a fireplace is most appropriate for people to whom the esthetic value of fire is paramount, and the heat it produces is of little or no importance.

Advantages

1.) *Looks.* What else is there to say?

2.) *Little creosote.* Fireplaces are clean-burning, at least in comparison to most old airtight woodstoves, but it is also true that some of the steps that you can take to make a fireplace a more efficient heater, such as installing glass doors and regulating the burn with an adjustable damper, will produce more creosote.

Disadvantages

1.) *Inefficiency.* What else is there to say?

2.) *Price.* This will vary a good deal, depending upon what sort of fireplace you want, but a good, mason-built Rumford will cost much more than a woodstove and will never amortize itself in any tangible way, except possibly by increasing the resale value of your house.

3.) *Maintenance.* This does not apply so much to masonry fireplaces (though they, too, need repair from time to time), as to factory-built fireplaces with metal fireboxes. The metal used in these units is fairly thin sheet steel, prone to rusting, warping, buckling and damper problems, all of which can be difficult and/or costly to repair.

MASONRY HEATERS

There are a number of different kinds of masonry heaters—Russian fireplace, Finnish contraflow and tiled heaters from Germany and Austria among them—which, despite differences in design and appearance, exemplify an approach to wood heating that differs sharply from that of stoves and furnaces. Instead of a round-the-clock fire maintained by periodic stoking and control of the air supply—the modus operandi applied to other serious wood heating equipment—masonry heaters rely on very hot fires—at times in excess of 2,000 degrees Fahrenheit—of short duration. Fires lasting only an hour or two heat a masonry mass weighing a ton or much more (FIGURE 1). The mass then radiates the stored heat for 12 to 24 hours, depending upon the weather. The extremely hot fires result in very clean burns.

Masonry heaters reportedly surpass conventional heaters in their heat-transfer efficiency ratings—how much of the produced heat ends up in your house instead of going up the chimney. A conventional stove operating at high combustion efficiency may send up to half its heat up the chimney. A well-designed masonry heater, on the other hand, stores and radiates something on the order of 80% of the heat it produces. It does this by directing the intensely hot gases through a series of channels in the masonry mass. By the time the exhaust reaches the top of the chimney, it is almost cool, having left its heat in the masonry. The smoke does not deposit creosote if the appliance is properly operated, because the fire is so hot that the tars and organic compounds are consumed in the firebox.

The various kinds of masonry heaters differ mainly in their heat-extraction systems: The Russian fireplace uses its chimney—built with horizontal passages and a top-mounted damper to hold in the heat when the fire is out—to trap and store the heat. The fire could be built in almost any fast-burning appliance—a kitchen or sauna stove, a fireplace, a water heater—attached to the chimney. The Finnish contraflow, on the other hand, is a self-contained firebox and storage unit that is vented into a conventional chimney, and is, as of this writing, considered to be the most efficient and readily available masonry heater.

As intriguing as their efficiency makes them, masonry heaters are made even more intriguing, at least to me, by the vintage of the technology, which originated in ancient Rome and saw widespread application in northern Europe and Russia in the 1700s.

Best Applications

Because masonry heaters are large (a contraflow's "footprint" might typically be 3-by-5 feet) and very heavy, installing one in an existing house will often prove to be an expensive proposition, assuming that it's possible at all. A new home, however, can be easily designed around a masonry heater. Sized to your house, a contraflow

FIGURE 1.
Masonry heater

or Russian fireplace should provide all the heat you need and, because of the gentle, even nature of the warmth (the surface temperature of a contraflow seldom exceeds 150 degrees Fahrenheit), will be more akin to central heating than space heating.

Advantages

1.) *Beauty.* Masonry heaters, faced with stone, brick, tile or other masonry materials, provide a full measure of esthetic enjoyment. If you wish, you can have one built with a window so you can see the fire.

2.) *Utility.* A masonry heater is arguably the most effective and efficient wood heater you can buy.

3.) *Ethics.* Environmentally, masonry heaters evidently outshine all other woodburners and, perhaps, every heating system save solar: They produce a very small quantity of emissions and will use a quarter to half the wood that a stove would require to heat the same space.

4.) *Low maintenance.* Because masonry heaters are so clean-burning, your chimney will stay cleaner than it would with most woodburning appliances. Also, there is little in the system that is likely to wear out.

Disadvantages

1.) *Price.* Yes, indeed: A basic manufactured contraflow kit will cost in the vicinity of $3,000, and this does not include the chimney, the footing, the exterior finish or the labor. Scandinavian imports frequently cost $11,000 or more. Over a long period of time (maybe 20 years), however, a masonry heater should amortize itself, even when compared to the most efficient woodstoves, with the high initial cost being offset by reduced fuel and chimney cleaning requirements and the elimination of the need to replace catalytic converters.

2.) *Availability.* Clearly, you can't go to your local stove store, pick out the model of masonry stove you want, and bring it home in the trunk of your car. As of this writing there are masonry heater kits available from a Canadian firm, Temp-Cast 2000, of Port Colborne, Ontario, from the New Alberene Stone Company, of Schuyler, Virginia, and possibly from others. There are also a few masons who specialize in masonry heater construction. (Make sure you hire one who knows what he or she is doing: An incorrectly designed or constructed heater is costly and useless at best, and costly and dangerous at worst.)

3.) *Convenience.* If you bridle at the thought of kindling one or two fires from scratch every day, a masonry heater will seem inconvenient. Also, a totally cold heater may take hours after it has been fired to produce significant heat.

PELLET STOVES

As concern over the environmental impact of woodburning has increased, so too has the stove industry's attention to mitigating that impact. Much attention has been directed at designing new stoves, but some has also been given to designing new fuel: the pellet. Pellets are generally made of waste wood but, depending upon the availability of raw materials, may be made from agricultural waste or cardboard. Pellets require special stoves equipped with electric-powered forced air intake, and this, combined with the fuel's low (5%) moisture content, leads to very low emissions and to overall efficiency ratings that are claimed to be a bit higher than those of catalytic stoves. Because the draft is forced and the exhaust gases are relatively cool, it is possible, depending upon the manufacturer's recommendations and local building codes, to vent a pellet stove with a relatively inexpensive double-walled pipe in lieu of a chimney.

At this point, more than half of the woodstoves being sold in the northwestern United States are pellet stoves, due in part to strict local emissions regulations and in part to the large number of pellet mills in the region. If you live elsewhere, make sure that pellets are available before you buy a stove.

Best Applications

People who live in areas unsuited by climate and/or topography to conventional woodburning equipment may find pellet stoves appropriate, as will people without ready access to firewood. Pellet stoves will also appeal to those seeking convenience: The operator must reload the hopper from time to time, but the stoves are otherwise self-stoking.

Advantages

1.) *Ethics.* Pellet stoves produce little in the way of emissions and burn fuel that would otherwise probably be waste.

2.) *Looks.* Models with glass doors allow you to see the glow of the flames.

3.) *Cheap and simple installation.* A stove that doesn't require a conventional chimney can save you a bundle.

Disadvantages

1.) *Price.* Pellet stoves often cost more than $2,000, and the fuel, at $150-$200 or more per ton (equivalent to 1½ cords of good hardwood), may be more expensive than cordwood in your area.

2.) *Dependence on electricity.* When the power goes off, so does the pellet stove, since it loses its forced draft. I can live with candles and flashlights, unwashed dishes and unflushed toilets, but I insist upon being warm. If you live in an area subject to power outages, you might want to think twice before you buy.

3.) *Availability of fuel.* Again, if you live in an area with a sufficiency of pellet mills, no problem. In my area, we had one, but it shut down last year, leaving the people who bought stoves with the choice of paying a premium for fuel trucked in from afar, or abandoning their expensive purchase.

GAS STOVES

Now we're really straying far afield. There are, however, gas stoves that look like woodstoves, cost about the same as woodstoves, produce about the same amount of heat as woodstoves and allow you to see the fire, as do many woodstoves. If you are going to heat your home with fossil fuels, I suppose this is a pleasant alternative—or, at least, supplement—to the soulless appliance in the basement.

Best Applications

People who want their heat to come with a glow but no hassle or mess will probably like gas stoves, as will people who are not physically able to cope with wood or coal, but who want a cheery space heater.

Advantages

1.) *Convenience.* Just pay the monthly gas bill. How hard is that?

2.) *Looks,* already discussed.

3.) *Availability of fuel.* Models are, or will be, available that burn either propane or natural gas, so you should have no problem finding a supplier. Until we run out, anyway.

Disadvantages

1.) *Dependence upon fuel supplier.*

2.) *Fuel cost.* It will generally cost more to heat a space with a gas stove than with a woodstove.

3.) *Possible chimney damage.* As gas stoves are improved, as they surely will be, and their combustion efficiency exceeds 75%, as it may, people living in cold climates may find that conventional masonry chimneys deteriorate due to acid condensation from the cool flue gases produced by gas appliances.

STOVE INSTALLATION

I would guess that it's a rare chimney sweep who doesn't encounter a large number of stoves that are, for one reason or another, unsafely installed. My own very unscientific survey indicates that 75% of the installations I see for the first time pose significant hazards to the people who share the house with them. Solving these problems is sometimes as simple as securing the joints of a stovepipe with sheet metal screws, but often is far more difficult. Old houses with old chimneys and stoves are frequent offenders, but new houses are far from immune.

In most cases, unsafe stove installations are dangerous because the appliances are located too close to combustible materials. The kindling temperatures of walls, ceilings, floors and furniture that are too close to stoves and stovepipes are reduced over time. That means they can ignite more easily than objects located a safe distance from the stove. It's true that these hazards are theoretical, but it's also true that they sometimes realize their destructive potential outside the testing lab.

As a consumer, you should be able to depend upon a certified chimney sweep to recognize whether or not your stove is safely installed (though you should bear in mind that some unsafe conditions are hidden from view), and, perhaps, a knowledgeable insurance inspector will know what to look for. But a surprising number of people who should know the safety requirements—masons, building contractors and firefighters—often do not. If you have a stove or are contemplating having one installed, it therefore behooves you to know enough about safe stove installations to avoid dangerous mistakes and recognize when to get help.

Note that we've already discussed chimneys: I'm assuming here that yours is adequate. Also, all new stoves have, or should have, specific instructions in the owner's manual and on a tag attached to the stove concerning their recommended clearances and installation requirements. These instructions supersede anything that I say here, as do local and/or state building codes.

Clearances

Stoves. Unless the manufacturer's instructions say otherwise, a stove should be 36 inches from anything combustible (except for the floor—more on that later). This includes walls, drapes, chairs, ceilings, mantelpieces—anything that can burn. By installing a heat shield like the one shown at the head of this chapter—a sheet of noncombustible material such as sheet metal or masonry, spaced 1 inch from the combustible surface and open on the top, bottom and sides to allow free air circulation—the 36-inch clearance can be reduced to 15 inches. Take care, however, that the shield is large enough to cover any area within 36 inches of the stove. Some stoves have built-in heat shields which reduce the clearance necessary behind the stove. Install these stoves according to the manufacturer's specifications.

Stovepipe. Single-wall stovepipe should be 18 inches from any combustible surface. By installing a heat shield—either on the pipe or on the surface being protected—this distance can be reduced to 9 inches. There is double-wall stovepipe available that is approved for 6-inch clearance with 6-inch pipe (8-inch clearance is required for 8-inch, double-wall pipe), and can therefore solve some knotty installation problems. Be careful, though, that the reduced clearance the pipe makes possible doesn't bring your stove itself too close to anything combustible.

Understove Protection. Again, new stoves will be labeled with their clearance requirements, but in the absence of these, the protection, known as the stove board, should extend 18 inches beyond the stove in all directions. Its composition should be as follows: For stoves with legs 6 inches long or longer, the understove protection should be 2-inch-thick masonry (bricks, usually) covered with 24-gauge sheet metal. For stoves with legs 2-6 inches long, the masonry should be 4 inches thick, and hollow, and laid so that the hollows form continuous open-ended passages for air circulation. The sheet metal covering is the same as for the other board. Stoves with legs shorter than 2 inches must not be installed on combustible floors at all. Most stove dealers have or can get manufactured stove boards in various sizes, and half-inch Homosote NCFR wallboard covered with sheet metal is understood to provide adequate understove protection, though it is not specifically intended for this purpose.

Many otherwise careful people dismiss the need for special floor protection, saying something like, "Oh, my cat sleeps under the stove all the time. How hot can it be?" I would point out that the cat, unlike the floor, is mobile (though I've known tom cats who were barely so), and can get away from the stove and cool off. Also, a cat wouldn't begin to smolder without letting you know about it; a floor would.

Installing Stovepipe:

Besides the clearances, there are other things worth knowing about stovepipe:

1.) *Less is more.* Pipe releases a surprising amount of heat, in many installations as much as the stove itself. A large amount of pipe, therefore, encourages creosote formation by allowing the smoke in the pipe to cool before it reaches the chimney and by encouraging you to run the stove slowly, since you get so much heat from the pipe. The National Chimney Sweep Guild recommends not using more that 8 feet of pipe in an installation, unless it is unavoidable, in which case, the pipe should be supported every 4 feet with a solid bracket or hanger.

2.) *Corners.* Concerning 90-degree pipe elbows, less is more, also. Smoke is like a car—corners slow it down, encouraging cooling and creosote deposition, so use no more than two elbows, not including the chimney's thimble itself.

3.) *Stovepipe comes in different gauges:* The lower the number, the thicker the pipe. The NCSG recommends 24-gauge pipe for most stoves, and it certainly won't hurt to use 22-gauge. Pipe of these thicknesses stands up to daily use and to pipe fires much more reliably than does pipe in the lighter gauges 26 and 28, and will usually last enough longer to offset its higher price. Another good cautionary note from the guild: Don't use galvanized stovepipe in a woodstove installation, because it gives off zinc vapor, which is toxic, at temperatures of 750 degrees Fahrenheit and higher.

4.) *Horizontal sections* of pipe should, ideally, slope up toward the chimney at a rate of ¼-inch for every foot of length. They should never slope down. Also, the seams should be on the top or side of horizontal sections to prevent liquid creosote from dripping out.

5.) *Vertical pipe* should be installed with male (crimped) ends pointing down, again to prevent creosote dripping.

6.) *Joints.* Each pipe joint should be secured with at least three sheet metal screws.

Wall Pass-Throughs

We've discussed this in previous chapters, so I'll spare you further horror stories. Connecting a stove to a chimney through a combustible wall is tricky business. It's easy to be fooled by noncombustible facades such as brick or tile hearth walls, but you need to look closely, because if your chimney is on one side of a combustible wall and your stove is on the other, you likely have a problem. Very seldom, even in new houses, do I see the problem safely dealt with. As of this writing, the National Fire Protection Association and the NCSG recognize only five ways of connecting a stove to a chimney through a combustible wall:

1.) There are products—heavily insulated thimbles—that are approved for a wall pass-through if surrounded by a 2-inch clearance to combustibles.

2.) A regular clay tile thimble surrounded by 1 foot of solid masonry on all sides is safe. You rarely see this—though tile thimbles with much less than 1 foot of surrounding masonry are commonly used—because such a pass-through is large: approximately 32 inches square if a thimble with a 6-inch inside diameter is used.

3.) A section of factory-built, insulated metal chimney with a 2-inch air space surrounding it and a single-wall stovepipe installed through the center of the chimney section, with a 1-inch air space surrounding it (an 8-inch chimney section with a 6-inch pipe, for example) can be used, though centering the pipe is difficult (FIGURE 2).

4.) The same section of factory-built chimney with a 9-inch air space all around it may be safely used without the smaller pipe passing through it. Confusion is common because most factory-built metal chimneys are approved for a 2-inch air space, but these clearances are safe only when the product is part of a *vertical* chimney; as a wall pass-through device, it needs the 9-inch clearance. Note also that a chimney section used in this way should not be simply butted up against the masonry chimney; it should make a positive, tight connection with it (though it shouldn't actually project into the flue).

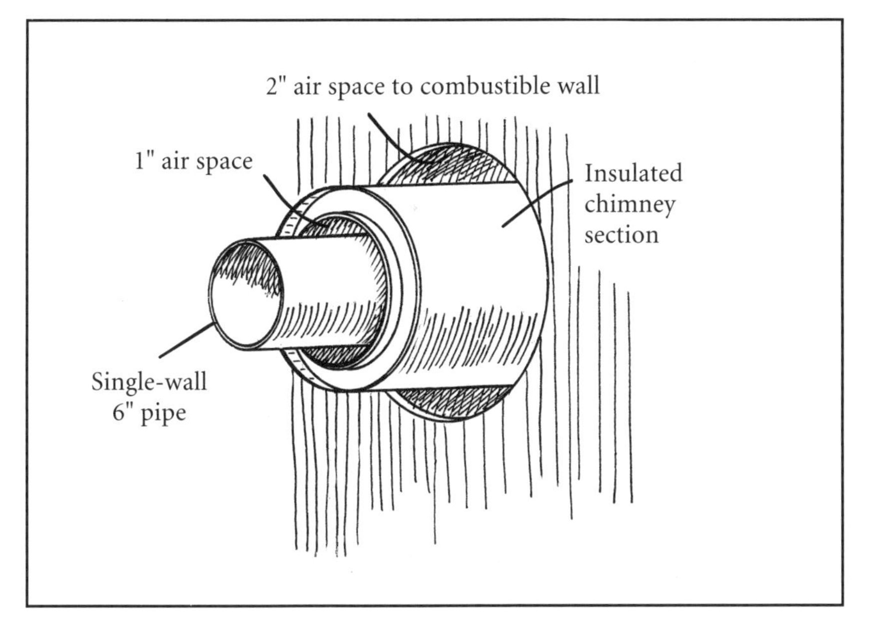

FIGURE 2.
Wall pass-through with stovepipe inside chimney section

5.) A ventilated metal thimble with two (not one) 1-inch air passages completely surrounded by 6 inches of noncombustible insulation is acceptable, but since, as of this writing, no such device is commercially available, this option should be disregarded.

Fireplace Installations

Many houses built after central heating became commonplace have fireplaces, but it wasn't until after the oil crisis of the early 1970s that woodstove hookups to fireplaces became commonplace in new construction. When the oil crunch struck, many people bought woodstoves and, in one way or another, installed them in their fireplaces. There were two commonly used techniques for doing this:

1.) Stovepipe was connected to the stove and extended just past the fireplace damper, and the remaining space in the damper opening was filled with fiberglass insulation or a sheet-metal plate.

2.) A metal plate that covered the fireplace opening was installed and the stovepipe passed through a hole in its center. Several manufactured stoves had this plate built in.

The problem with both installations is that smoke from the stove doesn't go directly into the chimney flue but lingers and cools in either the fireplace or its smoke chamber. When coupled with fireplace flues that are usually too large for a woodstove, the result is often a large buildup of creosote, particularly glazed creosote, in the chimney and fireplace. I vividly recall removing over 40 gallons of creosote from such an installation ten years ago.

A large fireplace flue should be lined to downsize it for a woodstove, and it should make a direct connection to the stove, but many people resist doing this, either because of the expense involved, or because the smaller flue makes it impossible to use the fireplace as a fireplace. Fortunately, there are ways of installing a stove in a fireplace which, while not addressing the potential problem of the flue being too large, do result in the smoke entering the flue without cooling in the fireplace or smoke chamber and, therefore, reduce the risk of dangerous creosote buildup.

Often the cheapest and, in my opinion, the best method is to install a thimble that enters the chimney flue, usually over the mantel. The damper is then shut and sealed with high-temperature silicone, or replaced with sheet metal if the damper is too warped to effect an airtight seal. The main drawback to this type of installation is that wooden mantels, trim and paneling often present difficult clearance problems. A second drawback is that the installation is semipermanent: The fireplace should not be used unless the thimble is filled with masonry and mortar.

There are connector kits available—usually flexible stainless steel oval-shaped pipe—which can be pushed up through the damper opening to join the chimney flue. The unfilled space in the damper is filled, usually with sheet metal. The oval pipe has a round thimble to accept the stovepipe. The main drawback of this method is its cost: The kits usually cost several hundred dollars.

A third stove-in-fireplace installation technique is simply to extend stovepipe through the damper opening all the way to the beginning of the flue. The unfilled space in the damper opening is sealed with sheet metal cut out to accommodate the pipe and high-temperature silicone. The only problem with this method is that very few fireplaces will permit it, either because most damper openings are too small or because the line from the stove to the opening of the flue is too indirect, or both.

One important safety note concerning stove-in-fireplace installation: Manufactured metal fireplace and chimney units— known as "zero-clearance" fireplaces—are engineered and approved for use as open fireplaces, but cannot accept woodstove installations or, for that matter, after-market glass doors. Either modification could cause excessive heat buildup which could transfer to nearby combustible walls. If you have a manufactured fireplace, use it only as detailed in the manufacturer's instructions.

STOVE OPERATION TIPS

Let me begin this discussion of how to operate woodstoves, furnaces and fireplaces with a large disclaimer: Nothing that I say is intended to supersede the instructions and advice found in your appliance's owner's manual, or the advice you receive from a knowledgeable stove dealer, chimney sweep or, for that matter, yourself. In short, the advice that I offer is purely generic, as it must be, since I have no specific experience with your particular heating system. I speak primarily to woodstove owners, with whom I group those who have wood furnaces; closed fireboxes and limited oxygen supplies have a greater potential to contribute to operational problems than do open fireplaces.

After years of burning wood myself and observing and dealing with other people doing the same, I have come to think of the pursuit as being akin to tightrope walking, balance being the element crucial to the success of either endeavor. The balance I'm speaking of is between the cleanest possible burn and the burn that provides you with the greatest amount of usable heat and therefore uses the smallest possible amount of fuel. Absent the technological assistance of catalytic converters and effective secondary combustion, the tightrope is a stark delineation: Slip off on one side and your chimney stays clean, your fuel consumption increases, and your stove will neither hold a fire long enough for convenience's sake, nor, in all likelihood, adequately heat your living space without constant refueling and overheating. Slip off on the other side of the tightrope, and you have long-lasting fires that contribute much of their heat to your home, but are smoky and lead to a dangerously dirty stovepipe and chimney.

For people with catalytic or high-tech stoves, the tightrope becomes less distinct but doesn't disappear. These stoves, it is true, consume part of the smoke produced by the fire, but miraculous as they seem, they still require oxygen for combustion: Too little will result in inefficient burning, and too much will result in inefficient heating. A catalytic or high-tech stove may make the best of a bad situation, but operated too far from the tightrope—on either side—neither will perform as it should.

You are probably tired of hearing me carry on about the interrelatedness of all the elements in a wood heating system, so I'll reassure you that, for the purposes of this discussion, I'm assuming that your stove, house and chimney are well matched to one another, the weather is cold enough to make woodburning sensible, and your wood is properly seasoned. This leaves you, the operator, and the critical variables that govern combustion and over which you have immediate control: heat, oxygen supply and fuel.

Complete or, at least, good combustion occurs in three stages:

1.) *Drying,* during which enough moisture is cooked out of the wood to allow it to burn;

2.) *Pyrolysis,* during which the wood's molecules are changed by exposure to heat, producing tar droplets and gases;

3.) *The charcoal stage,* which is the combustion that occurs after the tars and gases have either burned or exited the firebox.

The three stages can and often do take place simultaneously in different parts of your stove, or even in different places on the same piece of wood. All three of the stages, indeed the entire combustion process, require heat and oxygen; deprive a fire of either at any point in the combustion process, and it will soon go out.

Heat

Strike a match and you have heat, but not enough heat to drive the moisture from a firebox full of wood. This is why you start a fire with paper and kindling and gradually add larger pieces of wood only after the kindling and succeeding sticks are burning well: Wood in the first stage of combustion doesn't produce enough heat to drive the moisture from additional wood added prematurely to the fire. By not overloading the firebox, you assure that there is always wood in the heat-producing second and third stages of combustion.

Oxygen

Throughout combustion, the temperature of the fire is determined in part by how much oxygen is available to it. A fresh load of wood in the stove requires, for a short time, as nearly unlimited a supply of oxygen as the stove's design allows in order to evaporate the moisture in the wood and proceed to the next stage of combustion, at which point you are again on the tightrope. The fuel in the pyrolysis stage produces heat: Continuing to supply oxygen in unlimited or large amounts will, at least, make you uncomfortably warm for a little while and rapidly burn up the load of wood, and, at worst, result in serious overheating and damage to your stove and/or chimney. Restricting the oxygen supply too much, on the other hand, will prevent the gases and tars produced in this stage from burning, which, of course, will result in creosote formation.

How can you tell if you are adding enough oxygen to the fire? First, by installing a stovepipe thermometer and learning to understand it: The surface temperature it measures is approximately half the flue gas temperature. Flue gas temperatures below 400 degrees Fahrenheit indicate inadequate oxygen (assuming that there is sufficient fuel in the firebox for a fire), and flue gas that is hotter that 900 degrees Fahrenheit is potentially damaging to your chimney. Your stovepipe thermometer, then, should read between 250 and 475 degrees through most of the burn cycle.

A second good indication of how well managed your fire is will be the smoke coming out of the chimney once the fire is well established. Fires that have just been kindled usually produce a good deal of smoke for a short time. Thick gray, brown or black smoke is evidence of an oxygen-starved fire; light gray, almost white smoke with a good degree of transparency indicates that you are supplying enough oxygen for efficient combustion. It is often almost impossible to detect smoke coming out of the chimney of a well-run catalytic or high-tech stove except in exceedingly frigid weather. It was, in fact, my amazement at seeing one of those "smokeless" chimneys that convinced me to get a catalytic stove.

A third way of judging whether your fire is getting the right amount of oxygen is to look at it: Once it's past the first stage, the fire should be active, not smoldering (which indicates too little oxygen) or roaring (which indicates too much). Note that, if your stove doesn't have a viewing window, it may be difficult to assess the fire this way, since opening the door gives it a sudden supply of oxygen it doesn't have during normal operation.

Fuel

Fuel—in this case, wood—is the third ingredient necessary for combustion. In earlier chapters we discussed what kinds of wood are good, how to season it and how to store it, so what's left? How to put it in the stove is the answer, and I don't mean just opening the door and tossing it in. How you manage the stoking of your stove will have a large impact upon how effectively it heats and how clean your chimney stays. Consider the opposite extremes of stove stoking techniques, the rationales behind them, and some of their possible repercussions:

Woodburner A (not his real name) kindles a fire, fills the firebox as rapidly as possible, then lets the charge of wood burn entirely, or at least well into the charcoal stage, before repeating the process. He feels that he gets the cleanest possible burn this way because, 1.) He doesn't frequently open the stove door, thus introducing large amounts of room-temperature air and cooling the firebox; 2.) He isn't constantly adding room-temperature—or colder—wood, which will temporarily cool the fire; and 3.) He allows the fire to remain at its cleanest-burning stage—the charcoal stage—for a significant period of time. What he overlooks is that his chimney, particularly if it's exterior, may have a chance to cool between fuelings, increasing the possibility of creosote buildup. He also ignores the possibility that since he only adds wood when the fire is in a relatively inactive stage, he may unnecessarily prolong the first stage of combustion and encourage cool, smoldering fires.

Woodburner B only needs enough kindling each year for one fire—his first, because from then on he adds wood to the firebox while there are still active flames. B's rationale is that by maintaining a nearly constant, active fire, he'll keep his chimney warm enough so that it will be unreceptive to creosote deposits. Also, by adding wood to an active fire, B feels he encourages rapid evaporation of the moisture in the wood and prevents wide temperature swings. What he overlooks is that, 1.) Because he maintains a large load of fuel in his stove, he may have to overly limit its oxygen supply in order to avoid overheating; 2.) The overly full firebox that could result from his stoking procedure may inhibit the turbulence necessary to mix oxygen with the tars and gases given off by the wood during pyrolysis, resulting in incomplete combustion; and 3.) To a certain extent, the more wood you put in a stove, the more it will burn and the less completely it will burn it, perhaps because the entire fire is never maintained in the charcoal stage. If you've ever increased your feeding of a stove to counteract a cold snap, you may have seen the evidence of this in the large increase in quantity of ashes and unburned coals produced by the stove.

The stoking procedure that offers the best balance between the extremes and has the best chance of promoting a clean burn without excessive wood use or inconvenience is as follows:

1.) *After kindling the fire, never add more wood than the fire can handle.* Putting a large, unsplit piece of wood on a small pile of kindling will result in a smoldering burn. Better to add small to medium-sized pieces of wood gradually to maintain an active fire. Don't be too stingy, though: A fire is most active when there is fuel layered above it, rather than laid in one row on the floor of the stove. Pay attention to spacing the wood, too. As Daryle Thomas, of East Wallingford, Vermont, an expert on solid-fuel heating, says, "When placing wood in a stove, think of a successful marriage: Keep the logs close enough to feel the heat and far enough away to breathe."

2.) *Use the stove's draft controls.* When the firebox contains sufficient wood (half to two-thirds full is a good place to stop unless you're stoking it for the overnight burn, in which case you'll probably have to nearly fill it), adjust the air inlets to produce the desired temperature, flame and smoke conditions described previously.

3.) *Reload, starting with fairly small, split wood, when the previous charge has burned down to large, glowing coals.* Before adding wood, rake the coals toward the air inlet (usually on the loading door) so that they form a smooth, consistent bed covering the bottom of the stove.

4.) *For a burn of relatively long duration, add wood as described—a few pieces at a time—until the firebox is full.* Don't immediately close the air inlets; do it a bit at a time, and never close them all the way. You'll have to experiment to find the right setting for the air inlets, but recognize that your goal is not to have unburned wood eight hours after the final stoking; this would indicate that, for much of the burn cycle, there was too little oxygen present. Instead, be satisfied with a bed of coals adequate to start a new fire. Even taking the care described, a long-duration burn will probably produce creosote, but by only adding wood to an active fire and by not shutting the stove down until all the wood is actively involved in flames, you ought to minimize the buildup. Many people try to atone for slow, smoky overnight burns by "burning the stove off" in the morning, i.e., running the stove at a very high temperature for 15 or 20 minutes. This method is not nearly as effective as giving the stove adequate amounts of oxygen whenever it's burning, and is likely to start a chimney fire.

Catalytic Stoves

The above rules of operation are, for the most part, applicable to catalytic stoves, as well as to high-tech airtight and nonairtight stoves. The only significant difference is that, with a catalytic stove, you need to pay attention to properly preheating the catalyst before engaging it, since smoke passing through it will not ignite at temperatures lower than about 500 degrees Fahrenheit. This means that the temperature of the catalyst must be maintained at between 500-800 for 20 minutes after starting the fire. Most catalytic stoves have special probe thermometers that give accurate readings of the temperature in the catalyst. Follow the same preheating procedure for at least ten minutes after refueling if the temperature has dropped below 500.

High-Tech Stoves

High-tech stoves also require preheating to effect secondary combustion. Consult your owner's manual for specific instructions.

Fireplaces

As mentioned earlier, fireplaces are not as complicated to use as stoves can be, and unless yours has a problem related to design or construction, you either know how to use it already or will learn soon

FIGURE 3.
Laying a fireplace fire

enough without my help. I will, however, describe four practices that I've found helpful in starting and maintaining fires in fireplaces.

1.) Get the andirons out of the way. They may be decorative, but building a fire on them usually results in logs smoldering above a distant bed of coals once the kindling is burned up. Keep the fireplace grate, though, and plenty of ashes to protect the bottom of the firebox.

2.) To lay a fire, put a good-sized base log (3 or 4 inches in diameter) toward the back of the fireplace, crumpled newspaper balls in front of the log, and a log cabin of cross-hatched kindling (or a tipi, if you prefer) over the paper (FIGURE 3). When adding wood to the fire, add it so that it's cross-hatched with the ends of alternate rows of logs resting on the base log. This promotes good air circulation while allowing enough wood to stay in contact with the coals to maintain flames. And flames are what a fireplace is all about, aren't they?

3.) Preheat a flue with a slightly recalcitrant draft (one that tends to smoke a bit during a fire's early stages) by holding a flaming newspaper torch above the damper just before you light the fire. Actually, you aren't so much preheating the flue as you are getting warm air moving up and stimulating a draft. Don't burn your fingers.

4.) If your damper is adjustable, experiment with closing it part way as the fire burns low: If the draft is sufficient to keep smoke going up the chimney, you'll probably find that you get more heat in the room from the fireplace and burn a little less wood this way.

Sources

The following publications I have found useful generally, as well as in the preparation of this book:

Walter Hall. *Barnacle Parp's Chain Saw Guide.* Emmaus, PA: Rodale Press, 1977.

Workers' Compensation Board of British Columbia. *Fallers' and Buckers' Handbook.* Vancouver, B.C.: W.C.B. of B.C., 1978.

Donald Culross Peattie. *A Natural History of Trees of Eastern and Central North America.* New York, NY: Bonanza Books, 1954.

George W. D. Symonds. *The Tree Identification Book.* New York, NY: William Morrow, 1958.

Gerry Hawkes. *Cable Harvesting Systems for Small Timber.* Waterbury, VT: Vermont Department of Forests and Parks, 1979.

National Chimney Sweep Guild. *Successful Chimney Sweeping.* Olney, MD: 1987.

National Fire Protection Association. *NFPA 211, Chimneys, Fireplaces, Vents, and Solid Fuel Burning Appliances.* Quincy, MA: 1983.

Index